D1271152

Ions, Electrodes and Membranes
Second Edition

Frontispiece. A mosaic from Pompeii depicting the fight of an octopus with a lobster. The electric ray is shown at the top of the mosaic

Ions, Electrodes and Membranes
Second Edition

Jiří Koryta
Institute of Physiology,
Czechoslovak Academy of Sciences,
Prague, Czechoslovakia

JOHN WILEY & SONS

Chichester · New York · Brisbane · Toronto · Singapore

Other Wiley Editorial Offices

John Wiley & Sons, Inc., 605 Third Avenue,
New York, NY 10158-0012, USA

Jacaranda Wiley Ltd, G.P.O. Box 859, Brisbane,
Queensland 4001, Australia

John Wiley & Sons (Canada) Ltd, 22 Worcester Road,
Rexdale, Ontario M9W 1L1, Canada

John Wiley & Sons (SEA) Pte Ltd, 37 Jalan Pemimpin   05-04,
Block B, Union Industrial Building, Singapore 2057

Library of Congress Cataloging-in-Publication Data:

Koryta, Jiří.
    Ions, electrodes, and membranes   /   Jiří Koryta.—2nd ed.
        p.   cm.
    Includes bibliographical references and index.
    ISBN 0-471-93079-2 (cloth): —ISBN 0-471-93080-6 (paper)
        1. Electrochemistry.   I. Title.
    QD553.K673   1992
    541.3'7—dc20   91-20771
                                                            CIP

A catalogue record for this book is available from the British Library

ISBN 0 471 93079 2 (cloth)
ISBN 0 471 93080 6 (paper)

Typeset in Times 10/12pt by APS, Salisbury, Wiltshire
Printed in Great Britain by Biddles, Guildford, Surrey

Contents

CHAPTER 3   MEMBRANES   135

Preface to First Edition

The present book is an introduction to electrochemistry destined particularly for people coming from other fields of science. Electrochemistry is one of the oldest branches of physical chemistry and, in fact, of chemistry at all. The first great electrochemist was the Italian Luigi Galvani (1737–1798) who made his main discovery as early as 1791! In spite of this it took a considerable time, about three-quarters of a century, before two large regions formed in electrochemistry: one of them dealing with ion-containing liquids and the other one concerned with processes at electrodes. Later, particularly under the influence of the study of living cells, a new region emerged—the study of membranes. As I expect that at present this field has the greatest chance for important discoveries, I reserve a special chapter for it.

In view of its great age, electrochemistry has an entrenched kind of language which makes its study rather difficult, both for a biologist and a physicist. Therefore I have tried to approach each problem in a simple physical manner keeping in mind that electrochemistry is an experimental discipline. At the beginning I always describe a simple (sometimes rather idealized) experiment and only then formulate a more general law and discuss its consequences. In my opinion this is a more suitable way of explanation than to postulate a certain law of nature and only use the experiments to illustrate its consequences. Needless to say, the choice of detailed information shows how much I stress the importance of electrochemistry, particularly for biology.

The reader will find additional information on various topics in this book in the books and reviews quoted in the text and also in the textbooks on electrochemistry: D. A. McInnes, *The Principles of Electrochemistry*, Reinhold, New York, 1961; G. Kortüm, *Treatise on Electrochemistry*, Elsevier, Amsterdam, 1965; J. O'M. Bockris and A. K. N. Reddy, *Modern Electrochemistry*, Plenum, New York, 1970; and J. Koryta, J. Dvořák and V. Boháčková, *Electrochemistry*, Science Paperbacks, Chapman and Hall, London, 1973; and in three series of 'Advances', i.e. *Advances in Electrochemistry and Electrochemical Engineering* (eds. P. Delahay, H. Gerischer and C. W. Tobias), Wiley–Interscience, New York; *Modern Aspects of Electrochemistry* (eds. J. O'M. Bockris and B. E. Conway), Butterworths, London, and Plenum, New York; and *Electroanalytical Chemistry* (ed. A. J. Bard), Marcel Dekker, New York. Furthermore, there are three compendia of electrochemistry: C. A. Hempel (ed.), *The*

Encyclopedia of Electrochemistry, Reinhold Publishing Co., New York, 1964; A. J. Bard (ed.), *Electrochemistry of Elements* (a series in publication since 1973), Marcel Dekker, New York; and J. O'M. Bockris, B. E. Conway and E. Yeager (eds.), *Comprehensive Treatise of Electrochemistry* (a ten volume series in publication since 1980), Plenum Press, New York.

I am much indebted to Dr A. Ryvolová-Kejharová, Mrs D. Tůmová and my wife for their help with the manuscript and to Mrs Kozlová for drawing the figures. Dr M. Hyman-Štulíková kindly revised the English.

Preface to Second Edition

People working in physics, chemistry, biology and technology come into everyday contact with ions, electrodes and membranes. The present book is intended for beginners in all these fields with some preference for future biologists. There are no requirements imposed on the knowledge of the reader exceeding secondary school courses in science and, in general, the subject is treated more simply than in the first edition.

During recent years the pace of progress in the subject matter of the present book has been particularly rapid in the field of new electrically conducting materials, of membrane structure and of membrane transport, with important implications for neurophysiology and bioenergetics. In the present book, the beginner will find the basic facts and theories that will enable the chosen field to be followed.

The stimulating milieu of the Institute of Physiology of the Czechoslovak Academy of Sciences where I have been acting as a consultant since 1989 has considerably influenced the second edition of this book. I am much indebted to several scientists working in this Institute, particularly to Professor A. Kotyk, Dr F. Vyskočil and Dr (Mrs) V. Vlachová for their help. Professor Kotyk has made many helpful suggestions to the text, particularly in the field of membranology and biochemistry in general. Besides that he has taken pains to improve the English of the manuscript. Dr Vyskočil has made many useful comments on physiological topics. Dr Vlachová, a biomathematician and a gifted artist at the same time, has designed several cartoons to refresh the monotonous sequence of scientific information. Professor J. Dvořák has also made several valuable suggestions.

My thanks are also due to Mrs M. Kozlová, Mrs J. Pištorová and Mrs L. Kubínová for their technical help with preparation of the manuscript.

March 1991   Jiří Koryta

Introduction—Layman's Contact with Electrochemistry

A summer afternoon. Sitting in my garden I am pondering how to start this book. In front of me I see a green hedge which I prefer to an iron fence as it is nicer to look at and it does not become rusty. From time to time I grimace while eating sour cherries which are quite acid this year. The writing proceeds very slowly, as witnessed by checking my digital watch.

The reader may question the sense of these platitudes. In fact they directly concern the subject of the present book. The shrubs in the hedge are green since they contain chlorophyll which absorbs the sun's radiation. It is present in the leaves and stalks in tiny structures, chloroplasts, found inside the plant cells. The absorbed energy is then transformed into the chemical energy of sugar molecules in the process called photosynthesis. It takes place, for a substantial part, in thin membranes making up the so-called thylakoids which are stacked in the chloroplasts. Photosynthesis consists of a number of steps, one of which is an intermediate transformation of the energy of light into electrical energy displayed as an increase of voltage at those membranes.

Rusting of an iron fence is again a complicated process but to a lesser degree than photosynthesis. In a moist medium positively charged iron atoms, called ions, appear and readily react with water to form iron hydroxide. Negatively charged electrons remaining in the metal subsequently participate in a second reaction, namely the reduction of oxygen coming from the air to hydroxide ions, OH^-. This interplay of iron oxidation and oxygen reduction is the origin of rusting or corrosion, an unwelcome phenomenon causing multibillion dollar losses to the world economy every year.

Sour cherries are acid as they contain a rather large amount of hydrogen ions (protons combined with water to form oxonium ions, H_3O^+). These ions stimulate the taste buds in the tongue, inducing voltage changes at their membranes. This is the source of electrical current flow along nerve fibres which are connected to these taste bud cells. The final destination of these currents is the cortex of the brain where the electric impulses are transformed into new information, the sensation of acidity.

Finally, the digital watch! Its internal transistor circuit together with the digital display consume a small but not negligible amount of electric energy.

This energy is supplied by a tiny battery, a galvanic cell usually consisting of a mercury and a cadmium or zinc electrode. When electric current is supplied by the battery, oxidation and reduction processes take place at its electrodes, resulting in formation and disappearance of ions in an analogous way as during the corrosion of iron.

Among processes occurring in common life we find many that consist of reactions connected with the formation and decay of ions and with their movements, resulting very often in the appearance of electrical voltages at various structures. All these phenomena belong to the discipline that since 1814 has been called electrochemistry. In that year George John Singer published a book on the *Elements of Electricity and Electro-Chemistry*. What electrochemistry actually means will be found out by the reader when perusing the following pages. To give a definition of this concept without knowing its detailed content makes little sense, one reason being that some phenomena of electrochemistry also belong to the scope of physics (solid-state physics in particular), of biophysics, of membrane biology, etc. Be this as it may, all these branches of science should use identical or at least similar terms for the same concept.

Bibliography

With an exception of several classical monographs only books and reviews published after 1980 will be referred to in this and following sections.

Electrochemistry (including bioelectrochemistry and related subjects) has been dealt with in a number of textbooks, monographs, compendia and series of advances, which will be listed in this section. The references appearing at the end of each section will be concerned with the more detailed topics of that section.

TEXTBOOKS AND MONOGRAPHS

1. L. R. Faulkner and A. J. Bard, *Electrochemical Methods*, John Wiley & Sons, New York, 1980.
2. J. Koryta and J. Dvořák, *Principles of Electrochemistry*, John Wiley & Sons, Chichester, 1987.
3. P. H. Rieger, *Electrochemistry*, Prentice-Hall, Englewood Cliffs, N.J. 1987.
4. K. J. Vetter, *Electrochemical Kinetics*, Academic Press, New York, 1967.

COMPENDIA

5. *The Encyclopedia of Electrochemistry* (ed. C. A. Hempel), Reinhold, New York, 1961.
6. *Comprehensive Treatise of Electrochemistry* (eds. J. O'M. Bockris, B. E. Conway, E. Yeager *et al.*), 10 volumes, Plenum Press, 1980–1985.
7. *Electrochemistry of Elements* (ed. A. J. Bard), Marcel Dekker, New York, a multivolume series published since 1973.
8. D. B. Hibbert and A. M. James, *Dictionary of Electrochemistry*, Macmillan, London, 1984.

SERIES OF ADVANCES

9. *Advances in Electrochemistry and Electrochemical Engineering* (eds. P. Delahay, H. Gerischer and C. W. Tobias), Wiley–Interscience, New York, published since 1961.

10. *Electroanalytical Chemistry* (ed. A. J. Bard), Marcel Dekker, New York, published since 1966.
11. *Modern Aspects of Electrochemistry* (eds. J. O'M. Bockris, B. E. Conway *et al.*), Butterworths, London, later Plenum Press, New York, published since 1954.
12. *Advances in Electrochemical Science and Engineering* (eds. H. Gerischer and C. W. Tobias), VCH, Weinheim, since 1990.

Chapter 1
Ions

CHILD OF PHYSICS, CHEMISTRY AND BIOLOGY

The collection of various animals from the Mediterranean Sea shown on the frontispiece of this book is a reproduction of a Roman mosaic found in the debris of Pompeii destroyed by the eruption of Vesuvius in 79 AD and depicts a lobster's fight with an octopus. In the upper part of the mosaic, among other Mediterranean fish, swims an electric ray (*Torpedo*). This interesting fish (even contemporary holidaymakers may come into rather annoying contact with it) had, besides lightning, the most severe electric discharge known to the ancient period (the discharges from amber, or electron in Greek, were much weaker). The ability of Torpedo to stun other fish had already been described by Aristotle. This property, however, was not ascribed by ancient authors to electricity, but to some strange cold or to an unknown poison produced by the animal.

Late Renaissance biologists, Redi and Lorenzini (seventeenth century), dissected the electric ray and found in its body a strange stacked structure which was, according to present knowledge, the electric organ of the fish. They thought that Torpedo's stunning effect was caused by a rapid expulsion of tiny corpuscles ('*effluvia*') from the organ which acted as projectiles. Other authors of the same period, such as Borelli (1680) and Réaumur (1714), believed that a mechanical shock spreading through water was the physical cause of the stinging ray's action.

2

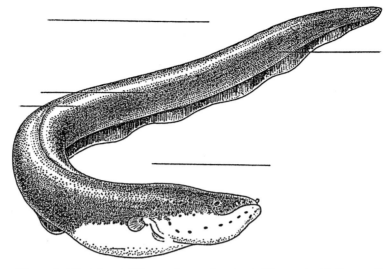

Figure 1. Electric eel (*Electrophorus electricus*). (Drawn by E. Opatrný)

Figure 2. According to a spectacular story South American Indians used horses to exhaust electric eels which could eventually be caught. (After E. Du Bois Reymond, 1849)

However, in the second half of the eighteenth century, Pieter van Musschenbroek, the discoverer of the Leyden jar, noticed a close similarity between the phenomena accompanying the discharge of the Leyden jar and the shocks elicited by *Torpedo*. The main objection of the supporters of the mechanical hypothesis against an explanation based on electricity arose from the absence of an electrical spark which usually accompanied such shocks. When comparing the discharge of the Leyden jar with Torpedo's shock the famous physicist Henry Cavendish came to the conclusion that the 'power' of the shock of the ray is not sufficient to induce a spark (the voltage of the discharge is, according to present knowledge, only 45 V). At the same time, the electric eel (*Electrophorus*) living in the Orinoco was brought to Europe; its electric effects were much stronger (see Figures 1 and 2). The English scientist and politician John Walsh was able to produce luminous sparks with this animal (its discharge voltage is about 600 V).

The following stage of research into 'animal electricity' was marked by the famous experiments of Luigi Galvani (1737-1798) commencing in 1786. In his first important result Galvani showed that the sartorius muscle of a frog's leg contracted when stimulated by a discharge from an electric machine. To Galvani's surprise the contraction also appeared in the case when the muscle and the nerve of the leg were simultaneously touched by a 'metallic arc' formed by two metals in contact (Figure 3). Galvani assumed that the electric effect had

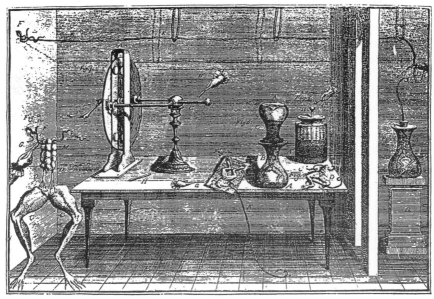

Figure 3. Galvani's electrophysiological experiments. As there has been a rapid decline in the number of frogs in the natural environment let us hope that nobody will again use them as experimental material

4

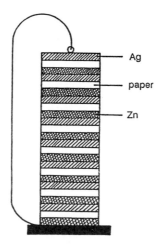

Figure 4. Volta's galvanic battery in its original version

its origin in the animal's brain, from where it was transferred through the nerve and the metallic arc to the muscle. When he informed Alessandro Volta (1745–1827) about his results Volta initially agreed with his views but later wrote to him that the observed electricity does not come from the brain but from the contact of the two metals. Needless to say, in his investigations Volta had progressed much further than Galvani. The structure of the electric organ of the electric fish stimulated his design of the voltaic pile (see Figure 4), which is not a very suitable term to use as in Italian *pila* means 'pillar, heap or tub'. Thus, we should rather call this device a battery. Before the invention of the dynamo by W. von Siemens (1816–1892) in the second half of the nineteenth century Volta's battery was the only source of electric current (strangely enough, called 'galvanic electricity' in contrast to 'voltaic electricity' produced by electric machines). Figure 5 shows a huge battery of Volta's cells which Sir Humphry Davy (1778–1829) had built in the cellar of the Royal Institution in London.

However, even Volta's contact theory of galvanic electricity could not correctly describe the reality. As late as 1835, Michael Faraday (1791–1867)

5

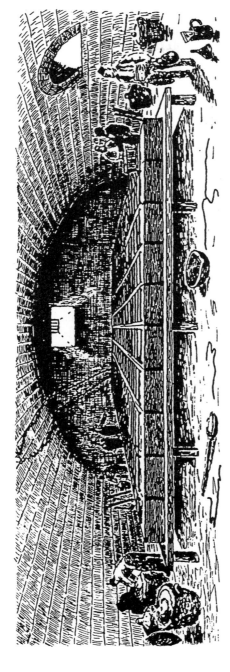

Figure 5. A gigantic voltaic battery placed by H. Davy in the cellar of the Royal Institution of Great Britain. (From a nineteenth century engraving drawn by J. D.)

6

proved that the origin of electrical 'energy'* was a chemical transformation in the cell.

References

1. L. Dunsch, *Geschichte der Elektrochemie*, Deutscher Verlag für Grundstoffindustrie, Leipzig, 1985.
2. W. Ostwald, *Die Entwicklung der Elektrochemie in gemeinverständlicher Darstellung*, Barth, Leipzig, 1910.
3. J. T. Stock and M. V. Orna (eds.), *Electrochemistry, Past and Present*, American Chemical Society, Washington, D.C., 1989.
4. C. H. Wu, 'Electric fish and the discovery of animal electricity', *Am. Scientist*, **72**, 598 (1984).

HOW IONS ARE FORMED

Let us proceed with Faraday who had devised a complete nomenclature of electrochemistry that was necessary for his research in *electrolysis* (also one of Faraday's terms). An electrically charged particle was called an *ion*, the positive particle was a *cation* while an *anion* was the negative particle. Faraday believed ions to be formed first during electrolysis whereas much later, in 1857, R. Clausius (1822–1888) conclusively proved that ions are present in an electrolyte solution regardless of electric current flow. Ions are not only found in an electrolyte solution but are met everywhere; they are truly ubiquitous.

Atoms are usually viewed as the basic building blocks of matter. However, this statement is not exact. There is no doubt that an atom is electroneutral as a whole since the positive charge of its nucleus is equal to the total negative charge of its electrons. However, one can find relatively few electroneutral atoms in our surroundings. Only rare gases (helium, neon, argon, crypton and xenon) are present in the atmosphere as unbound atoms. Another, much more frequent, component of the environment is the molecule; molecules are electroneutral only when viewed from the outside, but formation of chemical bonds within them causes considerable shifts of electrons between individual atoms to occur. In the form of electroneutral molecules, oxygen and nitrogen predominate in the atmosphere while water covers a major part of the Earth's surface. The solid portion is formed in part by molecules of quartz (silicon oxide) but silicates are also strongly represented. These are not electroneutral molecules but consist of negatively charged silicate anions and positively charged cations of aluminium, calcium, magnesium, sodium, etc. Within the biosphere, the situation is quite analogous. The principal component of all organisms is water. The walls of plant cells are built mainly of cellulose molecules, but equally important components like proteins, nucleic acids, inorganic salts, etc., are present inside

---

* The term 'energy' was used by Faraday as well as by his teacher Davy only symbolically, as the law of energy conservation was formulated much later by J. H. Meyer, but Faraday's unusual imagination had led him in the right direction.

cells in ionized form. Moreover, ions are found in the more distant regions of the universe. Silicates are present on the surface of the Moon and of Mars. A basic component of the sun is plasma, an extremely hot gas consisting mainly of protons and deuterons (in fact, light and heavy hydrogen ions), of helium ions and of electrons.

If all phenomena where ions participate were included in electrochemistry it would comprise a sizeable part of all science. Such a possibility—or danger? —could have arisen in the first half of the last century when Faraday included his discoveries in electromagnetism in his lectures on chemistry in addition to the phenomena connected with electrolysis!

A simple experiment (different from that which brought Faraday to the idea of ions) will demonstrate how ions are formed. Gaseous sodium can be obtained by evaporation of metallic sodium in an evacuated tube at a sufficiently high temperature (above 883 °C). The radiation from a light bulb is then passed through this gas and a spectroscope is used to observe the line absorption spectrum of sodium (Figure 6). Each line corresponds to the radiation energy necessary for transferring the outer electron of sodium to a higher energy level after absorbing a photon with an energy $h\nu$. The Planck constant, which is a universal physical constant, is denoted by h and the frequency of light waves corresponding to the line in the absorption spectrum by ν. Close examination of the spectrum reveals a conspicuous feature, i.e. the distances between the lines gradually become smaller in the direction of higher frequencies (or smaller wavelengths) and finally reach a limiting value, called the edge of the spectrum. The energy corresponding to the edge of the spectrum is just sufficient for tearing off the outer electron to form the sodium ion:

$$Na \longrightarrow Na^+ + e$$

This process is called ionization and can be brought about by a number of other mechanisms. For example, a high-energy photon can knock out one of the inner electrons of the sodium atom which then restores the stable structure with occupied K and L shells by regrouping the other electrons. The final structure is

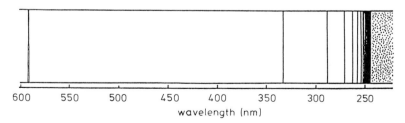

600   550   500   450   400   350   300   250
wavelength (nm)

Figure 6. The absorption spectrum of gaseous sodium. The adsorption lines are ordered according to their wavelength λ in nanometres. The frequency is $\nu = c/\lambda$, where c is the velocity of light in a vacuum. The line doublet on the left-hand side corresponds to the intense yellow colour in the emission spectrum of sodium. On the right-hand side the lines become denser until the edge of the series of lines is reached, corresponding to the ionization of sodium

again the Na^+ ion. Ionization can also be caused by collision of the atom with an elementary particle (electron, neutron, proton, etc.) or with other ions or atoms, etc. Ions are formed from molecules in the same way as from atoms. Negatively charged ions originate from reactions with an electron for example,

$$O_2 + e \longrightarrow O_2^-$$

There are simpler, more 'chemical' methods of forming ions. Consider feeding gaseous hydrogen chloride, which consists of HCl molecules, into water. The resulting electrically conducting solution of hydrochloric acid contains exclusively chloride ions, Cl^-, and hydrogen ions (which we shall provisionally denote as H^+) and virtually no HCl molecules. In contrast to pure water, this solution is an excellent conductor of electricity. During dissolution of hydrogen chloride in water the reaction

$$HCl \longrightarrow H^+ + Cl^-$$

obviously took place. Why could this happen so easily—at room temperature and without the influence of radiation? The answer can be found in the effect of the solvent. The solvent molecules form a stabilizing hydration sheath around each ion. The ionization of hydrogen chloride is followed and, in fact, made possible, by hydration of the ions formed from hydrogen chloride. In general, for an arbitrary solvent, this process is called solvation. The reaction of ionization of hydrogen chloride is accompanied by a further process:

$$H^+ + Cl^- + x\,H_2O \longrightarrow H^+(aq) + Cl^-(aq)$$

where the symbol aq indicates that the ions are solvated.

A similar electrically conductive solution consisting of hydrated ions is obtained by dissolving solid sodium chloride in water. This is somewhat different from dissolving hydrogen chloride in water. It has been proved by various physical measurements that the crystal lattice of sodium chloride consists of the sodium and chloride ions. The regular structure of rock salt, a typical ionic crystal (see Figure 7), is a result of electrical forces attracting the Na^+ cations and Cl^- anions and repulsing ions with the same sign (Cl^- and Cl^-, for example) and the electron sheaths of all the ions. This interaction determines the effective dimensions of the ions. The effective crystallographic

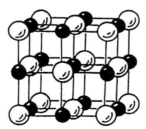

Figure 7. The structure of the cubic crystal lattice of sodium chloride. The sodium ions are represented by black spheres while the chloride ions correspond to white spheres

Table 1. Pauling's ionic radii in nanometres

			Li^+ 0.06	Be^{2+} 0.031		
O^{2-} 0.14	F^- 0.136	Na^+ 0.095	Mg^{2+} 0.065	Al^{3+} 0.05	Si^{4+} 0.041
S^{2-} 0.184	Cl^- 0.181	K^+ 0.133	Ca^{2+} 0.099	Sc^{3+} 0.081	Ti^{4+} 0.068
Se^{2-} 0.198	Br^- 0.195	Rb^+ 0.148	Sr^{2+} 0.113	Y^{3+} 0.093	Zr^{4+} 0.08
Te^{2-} 0.221	I^- 0.216	Cs^+ 0.169	Ba^{2+} 0.135	La^{3+} 0.115	Ce^{4+} 0.101

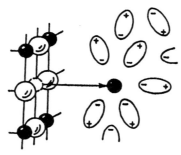

Figure 8. The sodium ion leaves the crystal lattice of NaCl and is solvated by water molecules

ionic radii given by Pauling are listed in Table 1. During dissolution the ions are only set free from the lattice and hydrated (see Figure 8).

POLAR MOLECULES

If two metal plates are placed parallel in a vacuum and connected to a d.c. source, then the electric current forms a positive charge on one of the plates and a negative charge on the other one, but no charge is transported between the plates—only the voltage (or, as it will be called here, the electric potential difference) increases. Such a device is called a condenser, since it helps the charge to become 'denser' on the plates.

The ratio of the charge brought to the positive plate of the condenser Q and the resulting electric potential difference is called the capacity of the condenser C:

$$C = \frac{Q}{\Delta V}$$

When the plates of the condenser are connected to a low-voltage source like an automobile storage battery (voltage 12 V) or to an electric grid (the a.c. voltage has to be first transformed to direct current by a rectifier), the condenser is charged instantaneously and the current flowing to the plates rapidly decreases. When the condenser is connected to a high-voltage source like a Whimshurst machine or van de Graaf generator, an electric discharge appears between the plates. The electrons jump from the negative to the positive plate and an electric current starts to flow.

In a subsequent experiment the space between the plates is filled with a substance having a high *electric dipole* like, for example, 1-chloropropane, $CH_3CH_2CH_2Cl$. Chlorine is an element with high electronegativity (the tendency to attract electrons); therefore the end of the molecule carrying a chlorine atom exhibits a slight excess of negative charge, while the other end becomes slightly positive. As a whole, the chloropropane molecule is electroneutral. Substances whose molecules have an electric dipole are termed *polar*.

The electric field that exists in the space between the plates of a charged condenser acts on the chloropropane molecules so that their positive ends are deflected somewhat towards the negative plate and the negative ends in the opposite direction. Complete orientation along the lines of force of the electric field is prevented by the thermal motion of the molecules—in a weak electric field the dipoles are aligned only slightly. The electric field also influences the distribution of charge inside some molecules to form new dipoles. The dipoles originally present in the molecule are called permanent while the new ones are induced. The dipoles partially compensate the electric effect of the charge present on the plates which results in an increase of the capacity of the condenser. Thus, at constant charge the voltage will decrease.

When the space between the plates is filled, for example with paraffin (which is a mixture of long-chain hydrocarbons), the capacity of the condenser increases very little. Such a substance is non-polar and the space inside the condenser behaves almost in the same way as in a vacuum.

Any type of matter that prevents an electric current from flowing between the plates of a condenser is called a *dielectric* (including a vacuum); the substances that have so far replaced the vacuum in the condenser are termed insulators which means that they do not conduct electric current.

The magnitude of a dipole is characterized by the *dipole moment*, the product of the positive charge at one end of the dipole and of the distance between the positive and the negative ends of the dipole. Dipole moments are determined for molecules in a gaseous state. In this aggregation the measurement is not complicated by mutual interactions of the dipoles. Some examples of dipole moments are listed in Table 2.

In the case of liquid or solid dielectrics their polar properties are preferably expressed by another quantity called *dielectric constant* or *permittivity*. The relative permittivity D defines the degree to which a given dielectric increases the capacity of the condenser compared with a vacuum condenser with the same size and distance between the plates. The values of the relative permittivities of some liquids are also listed in Table 2. The relative permittivity decreases with increasing temperature and is almost independent of the electric field (for weak fields).

Table 2. Dipole moments of gases and relative permittivities
of corresponding liquids

Substance	Dipole moment	Relative permittivity	
	μ^a	D	°C
Water	1.85	78.5	25
(Ice)	—	94	−2
Formamide	3.73	109	20
HCN	2.98	114	20
Formic acid	1.41	58	16
Acetic acid	1.74	6.15	20
Palmitic acid	—	2.3	71
Dimethylsulphoxide	3.96	48.9	25
Acetonitrile	3.92	37.5	20
Dimethylformamide	3.82	37.8	25
Nitrobenzene	4.22	34.8	25
Methanol	1.70	32.6	25
Ethanol	1.69	24.6	25
1-Butanol	1.66	17.5	25
tert-Butanol	—	2.7	25
Acetone	2.88	20.7	25
1,2-Dichlorobenzene	2.50	9.8	20
Chloroform	1.01	4.8	20
Benzene	0	2.28	20
Dioxane	0	2.28	20
Pentane	≈ 0	1.8	20.3
Ammonia	1.3	21	−34

^{a} Dipole moment unit is debye $= 3.336 \times 10^{-30}$ coulomb $\times$ metre.

The relative permittivity appears in the equation for the capacity of the plate condenser:

$$C = \frac{D\varepsilon_0 S}{d}$$

where ε_0 is the permittivity of a vacuum, $\varepsilon_0 = 8.85 \times 10^{-12}$ F m^{-1}, S is the area of one plate and d is the distance between the plates.

CONDUCTION OF ELECTRICITY

The effect of replacing the dielectric by a piece of metal touching both plates is not surprising. After the plates come into contact, an electric current starts to flow through the metal—the charge cannot be retained on the plates because they are conductively connected. Thus, metal is an *electric conductor*.

There are various kinds of conductors. They may be distinguished according to the type of electric particles that carry the current through them. In metals these particles are electrons—metals are electronic conductors. In the crystal lattice of a metal the metal atoms are ionized, the electrons are set free and the electron clouds* overlap in such a way that all the electrons are common to all the ions. Under these conditions the electrons in the metal behave in a way similar to molecules in a gas. (This follows from Fermi–Dirac statistics governing the behaviour of electrons in solids.) Thus, we speak of an 'electron gas' where the probable position of an electron changes in the same irregular manner as the position of a molecule in a gas, where it is constantly under the influence of collisions with other molecules.

Like electrons in an isolated atom, the electrons in a metal follow Pauli's principle which states that an energy level in the system can be filled by a maximum of two electrons with opposite spins. In a metal, the energy levels of the electrons differ very little from each other. Thus, they form a continuous band which we call the *conductivity band* (Figure 9).

The levels with lower energy are completely filled while those with higher energy are empty. The boundary between these two regions is not sharp but rather is continuous. The level which is just half-filled is called the *Fermi level ε_F*.

We already know that the electrons in the metal move in various directions and with various velocities. Connection of a voltage source to a metallic conductor will result in a slight change in the motion of the electrons—the component of the velocity in the direction of the field (i.e. from the negative to the positive pole) will prevail somewhat over the component in the opposite direction. This preference of the velocity in a certain direction is very small compared with the absolute magnitudes of the velocities of the irregular motion, but is just sufficient to cause the flow of an electric current. This is a typical property of many physical and chemical processes. They proceed in various directions with high velocities but in the equilibrium state all these 'partial' processes cancel each other out and no change is observed. When a partial process in one direction is slightly faster than in the opposite direction, a change in the physical or chemical system becomes apparent (current flow in the present case). Under these circumstances the rate of change (here the flow of electric charge, i.e. electric current) is proportional to the generalized force that causes the change (applied voltage or electric potential difference in the present case). Thus, we have deduced Ohm's law in a simple intuitive way, stating that electric current I is proportional to electric potential difference:

$$I \sim \Delta V$$

The proportionality constant is the conductance G which is equal to the reciprocal of the resistance R.

As mentioned above, the charge carriers in metallic conductors are electrons while metal ions in the lattice do not transfer electric charge because they occupy

---

* According to quantum mechanics the electron (as well as other elementary particles) cannot be considered as a particle with strictly determined position in space and velocity and direction of movement. Both these quantities are 'smeared out'. The electron cloud (the physicists call it 'orbital', which is an abbreviation of 'orbital function') characterizes the probability of appearance of the electron at a given position.

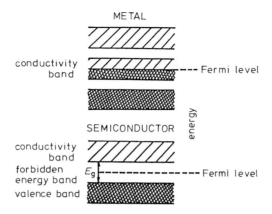

Figure 9. The band structure of a metal and of a semiconductor. Simply hatched areas are not occupied but can be occupied by electrons while void areas are not accessible to electrons. Double-hatched areas are occupied by electrons. The vertical coordinate of the scheme corresponds to the energy of the electrons. E_g denotes the energy difference between the upper and the lower edge of the forbidden energy gap (the gap width). In a metal, half of the conductivity band is filled with electrons. In a semiconductor, the conductivity band is empty. Between this and the valence band the forbidden energy gap is situated; the Fermi level lies in the middle of the gap. In comparison with a semiconductor the insulator has such a width of the forbidden band that the probability of electron jumping into the conductivity band is very small; thus the difference between a semiconductor and an insulator is quantitative only

fixed positions. These positions are not in fact completely fixed because ions vibrate around average lattice positions. The frequency as well as the amplitude of these vibrations increases when the temperature of the metal increases. The more the lattice ions vibrate, the more they hinder the free movement of the electrons, so that the resistance of a metallic conductor increases with increasing temperature.

When a piece of very pure elemental germanium is placed between two metal plates (the purification of such materials has been perfected during the past few decades) and a voltage source is connected to the plates, a very tiny electric current starts to flow. This current increases when the temperature is increased. In appearance, germanium is very like a metal, being grey with a metallic sheen, but it conducts electricity imperfectly. It belongs among the *semiconductors*.

The electronic structure of a semiconductor is different from that of a metal. The energies of the outer electrons in the electron sheath of the atoms in the semiconductor almost completely fill the *valence band* (Figure 9). Above this band lies the conductivity band, which is almost empty. There is no energy level that could be occupied by the electrons of the semiconductor in the energy interval between these two bands. This property is described by the appropriate term *forbidden energy gap*. The thermal vibration of the electrons can be so great that they can 'jump up' to the conductivity band and participate in the

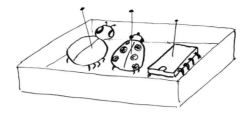

conduction of electricity. When the temperature is increased, more electrons reach the conductivity band, so that the conductivity of a semiconductor increases with increasing temperature. The electron that has 'jumped' up from the valence band has added a negative charge to the conductivity band. According to the law of conservation of charge, when a charge of a definite sign is formed, another charge of opposite sign must emerge at the same time. Thus, the electron has left a vacant position in the valence band—a *hole* or *vacancy*—with a positive charge. 'Chemically' speaking this means that an electron has been set free from the germanium atom and can move freely in the crystal lattice, while the resulting Ge^+ ion can annihilate its positive charge by an electron from the neighbouring atoms.

If an electric field is applied to pure germanium, the electrons migrate in one direction and the holes in the opposite direction (again the velocity of the electrons or holes in this particular direction slightly prevails over the velocities in the other directions). Such 'intrinsic' semiconductors conduct electric current poorly and are technologically relatively unimportant. A revolution in electronics, marked by the progress from simple rectifying diodes to microprocessors with integrated circuits, was based on the discovery of semiconductors with artificially implanted impurities, i.e. of semiconductors doped with minute quantities of admixtures.

A very small amount of arsenic can be added to pure germanium, leading to the following 'chemical' process: an electron is split off from the outer shell of the electron sheath of arsenic and is added to the neighbouring germanium atoms in such a way that it can move freely in the germanium lattice. Because of its ability to donate electrons, arsenic is termed an 'electron donor'. In the band scheme of germanium this phenomenon is demonstrated as shown in Figure 10. The admixture of arsenic brings about a new donor energy level just below the conductivity band. The electrons need only a small amount of energy E_d to move up to the conductivity band, thus increasing the conductivity of germanium. The semiconductor that contains a donor admixture and has conductivity based on the motion of electrons (negative charge carriers) is termed an *n*-type semiconductor.

The opposite process occurs when germanium is doped with boron instead of with arsenic. Boron is an electronegative element with a tendency to complete its electron shell. It receives one electron from germanium and is converted to the negative ion B^-. At the same time, a hole is formed in the electron structure of

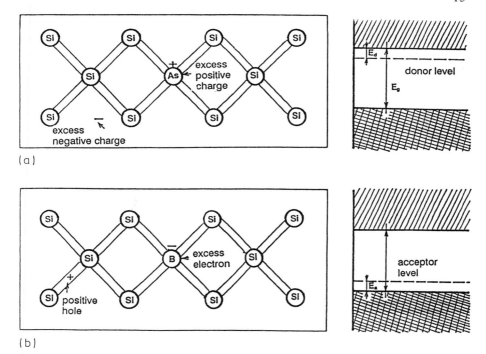

Figure 10. Schematic representation of a semiconductor (germanium) with impurities. Doping with an electropositive element (arsenic) results in a new 'donor' level immediately below the conductivity band. The electrons jumping from this level to the conductivity level induce n-type conductivity in the semiconductor (the charge carriers are the electrons with a negative charge). By doping with an electronegative element (boron) a new 'acceptor' level is formed just above the valence band. The electrons jumping from the valence band into this new band leave holes (positive charge carriers) in the valence band. Thus, the material becomes a p-type semiconductor. (From C. Kittel)

germanium. This situation is also depicted in Figure 10. Boron is an 'electron acceptor' and, when present as an admixture in germanium, gives rise to a new acceptor energy level just above the valence band. The transfer of an electron from the valence band to the acceptor band requires a small energy E_a. The hole remaining in the valence band contributes to the conductivity of the semiconductor. Because it is a positive charge carrier, this material is called a p-type semiconductor. Increasing the temperature has the same influence on a doped semiconductor as on intrinsic semiconductors, because the thermal energy of the electrons permits them to reach the donor or acceptor levels.

In the next experiment, two plates made of a certain metal, for example cadmium, are immersed in a cadmium sulphate solution. When a source of d.c. voltage is connected to the plates, an electric current starts to flow, accompanied by chemical changes at the cadmium plates, which will be termed *electrodes*.

The cadmium plate through which the positive charge from an external source of electricity enters the solution is dissolved. This process can be described by the equation

$$Cd(metal) \longrightarrow Cd^{2+}(solution) + 2\,e(metal) \qquad (1)$$

Metallic cadmium is deposited on the other electrode, where the electrons enter the solution, apparently in the process

$$Cd^{2+}(solution) + 2\,e(metal) \longrightarrow Cd(metal) \qquad (2)$$

The electrode that accepts a positive charge from the solution or injects a negative charge into the solution is termed the *cathode*, while the other electrode injecting a positive charge into the solution or accepting a negative charge from the solution is the *anode*.

Close examination of equations (1) and (2) reveals that the substances present in the solution which are formed at the electrodes as a result of the flow of electric current through the solution, or from which the final products originate, are exclusively ions. We can assume (and other experiments have justified our doing so) that only ions participate in the conduction of electric current in the cadmium sulphate solution or, in other words, only ions function as *charge carriers*. The conductor where electricity is communicated by the ions is termed an *electrolyte*.

The whole process described above is called electrolysis. This term, as well as electrode, anode, cathode and electrolyte, were also introduced by M. Faraday in 1835.

Electrolysis can be separated into two principal processes:

1. The processes taking place directly at the electrodes which are connected with charge transfer between the electrode and the electrolyte solution.
2. Charge transfer in the bulk of the solution called ion migration (Latin *migro* = wander). In addition, matter is also transported by other processes in the solution (see page 58). The processes taking place directly at the electrodes are dealt with in Chapter 2.

The ions that are present in the cadmium sulphate solution are Cd^{2+} cations and SO_4^{2-} anions. At higher concentrations $Cd(SO_4)_2^{2-}$ ions become noticeable, but will be neglected here as the concentration used in our experiment will not be high.

Imagine that the charge in the electrolyte is transferred only by the motion of the Cd^{2+} cations and that the SO_4^{2-} anions remain in fixed positions. Then, during current flow, the $CdSO_4$ concentration would remain constant at any point in the solution. However, in an actual experiment, the concentration around the electrodes increases at the anode and decreases at the cathode as a result of the mobility of both ionic species in the solution (in solution the Cd^{2+} cation moves more slowly than the SO_4^{2-} anion, but this is not important here). Therefore, when the Cd^{2+} ions are discharged at the electrode, new cadmium ions migrate to the electrode and, at the same time, sulphate ions migrate from the electrode. However, the electroneutrality condition must be preserved in the solution so that the total charge of the cations and the anions must be equal in any, even very small, volume of the solution. If this condition were not observed in a region of the solution, a space charge would appear and the repulsing force between the ions would expel excess ions from that region. The electroneutrality condition is expressed mathematically by the equation

$$z_+ n_+ + z_- n_- = 0$$

where z_+ and z_- are the charge numbers of the cation and of the anion, respectively, and n_+ and n_- are the numbers of cations and anions, respectively, in a volume of the solution (the exact definition of concentration can be found on page 31). The charge number is defined as the ratio of the charge of the ion to the proton charge (for example, for $CdSO_4$, $z_+ = 2$, $z_- = -2$). Thus, when the sulphate ions migrate from the neighbourhood of the cathode, the total concentration of cadmium sulphate there must decrease in order to preserve electroneutrality. On the other hand, the concentration of cadmium ions at the anode increases as a result of anodic dissolution of cadmium. Those ions migrate from the electrode and new sulphate ions approach it. In this way the electroneutrality is retained, leading to an increase in the cadmium sulphate concentration in the neighbourhood of this electrode.

The same properties as those exhibited by a cadmium sulphate solution are characteristic for solutions of innumerable salts, acids and bases (see page 38). However, the processes taking place directly at the electrodes are often quite complicated and differ considerably from those in the bulk of the solution, where ion transport proceeds. Fortunately, it is possible to separate the processes at the electrodes from those taking place 'inside' the electrolyte solution and thus to elucidate both groups of phenomena. This separation was not apparent to the founders of electrochemistry such as Davy, Faraday, Crotthus and others in the first half of the nineteenth century, and their views reemerge with remarkable perseverance even now.

There are, of course, many more different types of electrolytes. For example, two metal plates connected to a d.c. voltage source can be fixed to a sodium chloride crystal. Although the crystal is composed of chloride and sodium ions, no current starts to flow. Lattice forces hold the ions in their positions and the electric field, if not extremely high, cannot remove them from these sites.

Thus, solid sodium chloride is a non-conductor (*insulator*). As a substance, though, it is also called an electrolyte because on dissolution in water it forms an electrolyte solution. Obviously, a substance that is an electrolyte does not immediately need to be a conductor, but must achieve this property after a certain procedure. Here, the sodium chloride will not be dissolved but the temperature will be raised above 800 °C. In this way we obtain an excellently conductive liquid, the *fused electrolyte* (or the *ionic liquid*). Fused electrolytes are rather important; for example, the slag in metallurgical processes is a silicate-type fused electrolyte and aluminium is produced by dissolving aluminium

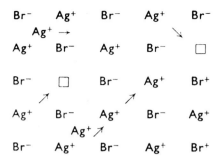

Figure 11. Charge transfer in solid AgBr. The conductivity is due to interstitial ions, Ag^+, and ionic holes. (According to J. I. Frenkel)

oxide (obtained from the mineral bauxite) in fused cryolite, Na_3AlF_6, and by electrolysis of this melt (page 113).

When, instead of sodium chloride, a crystal of silver chloride or bromide is placed between the metal plates,* we find that a distinct current starts to flow between the plates. This current is much smaller than that in a fused electrolyte or an aqueous electrolyte solution, but sufficiently distinguishes AgCl or AgBr from insulators like solid sodium chloride. Silver chloride or bromide are *solid electrolytes*. In their crystal lattice the halide (chloride or bromide) ions are immobile while the silver ions are genuine charge carriers. Because of imperfections of the crystal some silver ions lie outside regular lattice positions (they are at interstitial positions; see Figure 11) and some lattice positions are not occupied (these free positions are termed ionic vacancies or holes). The transfer of silver ions to and from interstitial positions and vacancies enables the transport of charge in the crystal. The conductivity of solid electrolytes rises as the temperature increases. Solid electrolytes are important for ion-selective electrodes (see Chapter 3, page 143) and for certain types of fuel cells (see Chapter 2, page 97).

For chalcogenides (sulphides, selenides, tellurides) of silver, lead, cadmium, etc., mixed conductivity is typical. In these materials the electric current is carried by ions as well as by electrons. The relative contributions of the two types of charge carriers depend, in addition to other features, on the deviation of the composition of the crystal from the stoichiometric component ratio (for example, if the Ag_2S crystal contains a small concentration of excess silver or sulphur). Another solid electrolyte, lanthanum trifluoride, consists of a rigid matrix of LaF_2^+ ions while 'molecular holes' between them can accommodate mobile fluoride ions. For application of LaF_3 single crystals (substances with a structure almost identical with an ideal crystal) occur in membranes of ion-selective electrodes (see page 143). This group of materials also includes a sodium aluminate species (the so-called β-alumina) with the empirical formula

---

* Silver plates must be used which are brazed to the crystal with a silver solder.

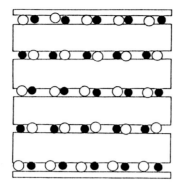

Figure 12. A scheme of β-alumina structure. The gaps between blocks of β-phase Al_2O_3 function as bridging layers sparsely populated by oxygen ($\bigcirc$) and sodium ($\bullet$) ions. (According to R. A. Huggins)

$Na_2O.11Al_2O_3$. At temperatures above 300 °C this material becomes rather conductive which is caused by mobile sodium ions situated between Al_2O_3 blocks (see Figure 12). Perspective application of this material is in the field of new storage batteries (page 96).

References

1. A. J. Dekker, *Solid State Physics*, Prentice-Hall, Englewood Cliffs, N.J., 1957.
2. J. Hladík (ed.), *Physics of Electrolytes*, Academic Press, New York, a multivolume series, published since 1972.
3. D. Inmann and D. G. Lovering (eds.), *Ionic Liquids*, Plenum Press, New York, 1981.
4. C. Kittel, *Introduction to Solid State Physics*, John Wiley & Sons, New York, 1976.

UNUSUAL CONDUCTORS OF ELECTRICITY

Recently new conducting materials have been discovered in the realm of polymer chemistry. In this way a new branch of electrochemistry, *electrochemical materials science*, is developing.* Existing or prospective applications to electrochemical technology (electrolytic cells, batteries, etc.) increase the interest in this area. Among these new substances, salts derived from a polymer of perfluorostyrenesulphonic acid should be mentioned first. The electroneutrality of the polymer is preserved by low-molecular-weight cations (*counterions* or *gegenions*, see also page 53), usually sodium ions. The structure of this substance, called NAFION by its originator, DuPont De Nemours Inc., is shown in Table 3, 1(a). Another polymer with ionic conductivity is polyethylene oxide (PEO)

---

* In this section compact polymer materials will be discussed while the properties of free polymer ions (polyelectrolytes) will be described later (page 50).

Table 3. Conductive polymers

1. Polymers with ionic conductivity (ion-exchanger polymers)

   (a) NAFION $+CF_2-CF_2+_x+CF-CF_2+$
   $$O[CF_2CF(CF_3)O]_zCF_2CF_2SO_3{}^-Na^+$$

   (b) PEO (polyethylene oxide)

   $$Li^+, ClO_4{}^-$$

   Li^+ is bound by ion–dipole interaction inside polymer helices

2. Polymers with electron conductivity

   (a) Electrons are delocalized inside the polymer chains in an analogous way as electron gas in a metal.

   Doped polyacetylene

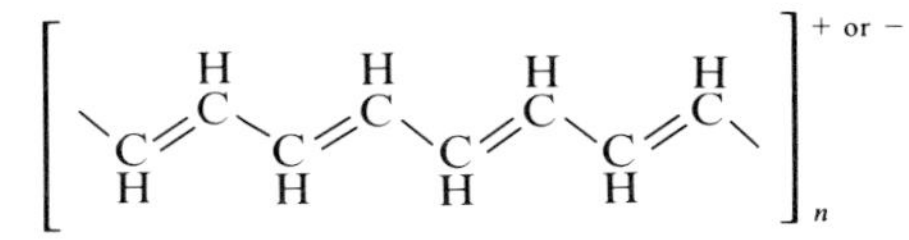

   Oxidized polypyrrole

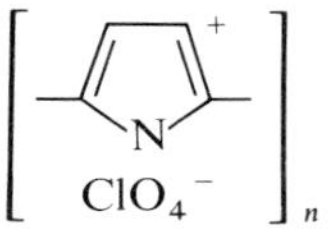

   (b) Polymer contains covalently bound redox groups.
   Polyvinylferrocene

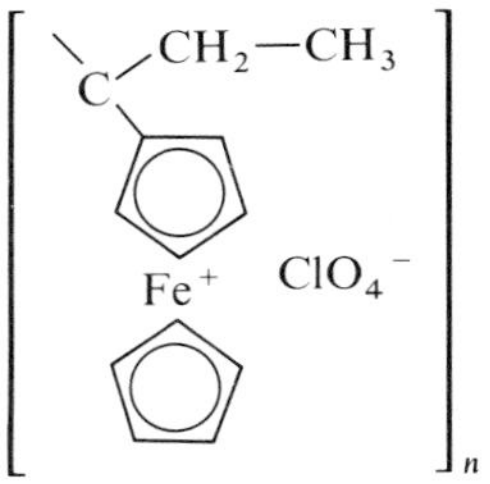

   Electrons are transferred by hopping.

   (c) Ion-exchanging polymer contains redox gegenions (e.g. methyl viologen in NAFION), which transfer electrons by diffusion.

---

containing a suitable salt, such as $LiClO_4$ (see Table 3, 1b). Lithium ions are bound to oxygen atoms of the polymer structure and perchlorate ions remain as mobile charge carriers.

Old textbooks of organic chemistry describe various reaction products of a resinous nature with a dark, even black, colour (e.g. aniline black), but it was only recently discovered that they can conduct electrons. Thus, polyacetylene doped with bromide or sodium ions (in fact, by the action of elemental bromine or sodium in order to inject charges into the polymer) possesses a very high conductivity so that the attribute, 'organic metal', is quite appropriate here. In this substance, the electrons of the double bond (the so-called π electrons) are not bound to a distinct pair of atoms ('localized') as one could conclude from the structure in Table 3, 2a, but they are shared by the whole polymer chain in an analogous way as the π electrons in the benzene molecule. These 'delocalized' electrons are similar to those of electron gas in a metal which is the cause of the rapid transfer of electrons or holes along the chain.

In another electronically conductive polymer the electrons preserve their individuality (e.g. in polyvinylferrocene; see Table 3, 2b), being present in electron-donor groups linked to the basic polymer matrix. From these groups they can jump to acceptor groups during electric current flow. The third type of electronically conducting polymers consists of a polymeric ion and of low-molecular-weight gegenions with electron-donor properties, such as methyl viologen (Table 3, 2c). In an electric field, they migrate across the polymer and transport electric charge injected by redox reactions (cf. page 81).

Good electron conductors include one non-metallic element, carbon, in its graphitic modification. In fact, graphite is a kind of polymer, a huge aromatic molecule with condensed benzene rings (Figure 13). When measuring the resistance of graphite single crystals (substances with a structure identical to an ideal crystal) a surprising result was found. Such a crystal conducted electricity very well in the direction parallel to the plane of the hexagonal net of carbon atoms while in the direction perpendicular to this plane there was no conduction at all (the property of solids that depends on the direction of measurement is called anisotropy, in contrast to isotropy which is a property independent of direction).

Strange solutions obtained by dissolving alkali metals, for example metallic sodium, in liquid ammonia will now be considered. The blue liquid deepens in

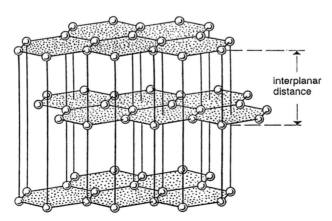

Figure 13. Graphite consists of stratified giant molecules formed of condensed benzene rings. (According to A. R. Ubbelohde)

22

colour when the concentration of sodium is increased, and above 4 mol Na l^{-1} the liquid becomes grey with a metallic lustre. Initially, the conductivity increases roughly in direct proportion to the sodium concentration, and when the liquid acquires a metallic appearance it increases rapidly (see Figure 14). The peculiar behaviour of these solutions is a result of the formation of solvated electrons in the liquid:

$$\text{Na(metal)} + \text{solvent} \longrightarrow e(\text{solution}) + \text{Na}^+(\text{solution})$$

The electrons are located in cavities between the solvent molecules and are very reactive (for example they act as strong reductants in many organic reactions). When their concentration increases, electron pairs, 'dimers', are formed. They possess antiparallel spins (the spin of one electron lies in the opposite direction to that of the other one). At still higher concentrations the orbitals of the electrons start to overlap in the same way as in a metal lattice. The electrons in a concentrated solution then form an electron gas, manifested by the change in appearance as well as by the striking increase in the conductivity.

The *high-temperature superconductivity* discovered by G. Bednorz and A. Müller was a great success in 1986 science. The low-temperature conductivity found in 1911 by H. Kamerlingh-Onnes (1853-1926) with some metals and alloys at temperatures near absolute zero is connected with the transition of these substances to the state of unbelievably low electric resistance. Bednorz-Müller superconductivity was acquired by a completely different group of substances at temperatures just below 30 K but even much higher temperatures were supposedly found as the upper limit of superconductive behaviour. The basic species for superconductive materials is the dual oxide La_2CuO_4 which is doped by lowering the oxygen content (the resulting compound, La_2CuO_{4-x}, contains a mixture of univalent and divalent copper) or by partial substitution of lanthanum with other elements so that the composition of the final compound can be formulated as $(Ba, Sr, Y)_xLa_{2-x}CuO_4$ (Figure 15) or simultaneously by a combination of both formulas. Remarkably enough, the discovery of high-temperature superconductivity disagreed with the accepted theory of

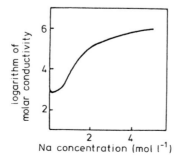

Figure 14. Dependence of the molar conductivity of sodium dissolved in liquid ammonia on the sodium concentration at $-34\,°C$. (According to E. C. Evers)

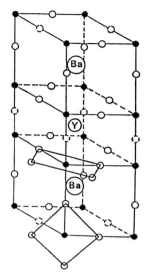

Figure 15. A high-temperature superconductor composed of the ions Ba^{2+}, Y^{3+}, Cu^{+} or Cu^{2+} (●) and O^{2-} (○). The semihatched circle is the crystal hole (vacancy). (According to D. C. Harris, M. E. Hills and T. A. Hewstone)

superconductivity worked out by J. Bardeen, L. N. Cooper and J. F. Schrieffer (Nobel Prize 1972).

References

1. C. E. D. Chidsey and R. W. Murray, 'Electroactive polymers and macromolecular electronics', *Science*, **231**, 25 1986.
2. J. Heinze, 'Electronically conducting polymers', in *Electrochemistry* (ed. E. Steckhan), Vol. IV, Springer, Berlin, 1989.
3. R. G. Linford (ed.), *Electrochemical Science and Technology of Polymers*, Vol. 1, Elsevier Applied Science Publishers, Essex, 1987.
4. C. P. Poole, T. Datta and H. A. Farach, *Copper Oxide Superconductors*, John Wiley & Sons, Chichester, 1989.
5. T. A. Skotheim (ed.), *Handbook of Conductive Polymers*, Vols. I and II, Marcel Dekker, New York, 1986.
6. A. R. Ubbelohde, *Endeavour*, **24**, 63 1965.
7. See also page xiv, Ref. 12, Vol. 1.

ION SOLVATION

A small quantity of sodium chloride can be dissolved in water to form a very dilute solution. Compared to the original situation with crystals of sodium chloride and water (at the same temperature), the solution is somewhat cooler.

When the same amount of sodium chloride is dissolved in methanol so that the resulting solution has the same concentration, the cooling effect is greater.

To elucidate these phenomena, we shall resolve the whole process into several steps characterized by the different amounts of energy accepted or released by the system. We are justified in doing so because this will be a thermodynamic procedure following the Hess law. This law states that the resultant energy change in a given system is independent of the steps through which the system proceeds in passing from the original to the final state. The following imaginary process can be considered (see Figure 16). First the sodium chloride will be evaporated into a vacuum. This step, connected with release of sodium chloride from the crystal lattice, requires the supply of the sublimation energy ΔH_{subl}.

All energy quantities appearing in this discussion are the *enthalpies*, i.e. the amount of heat supplied to or released from a system in a process taking place at a constant pressure. The symbol for the enthalpy is H. By convention, the energy accepted by the system is considered positive and the energy released negative.

The resulting gaseous sodium chloride will then dissociate into gaseous atomic sodium and chlorine. For this step the system must accept the necessary dissociation energy ΔH_{diss}. Then these gaseous atoms are transformed into ions:

$$Na(gaseous) + Cl(gaseous) \longrightarrow Na^+(gaseous) + Cl^-(gaseous)$$

Here the energy change ΔH_i is given by the sum of the ionization potential of sodium I_{Na} and of the enthalpy of electron acceptance by chlorine $\Delta H_{e, Cl}$:

$$\Delta H_i = I_{Na} + \Delta H_{e, Cl}$$

(The enthalpy of electron acceptance is sometimes called the electron affinity but, since in thermodynamics the term affinity is reserved for another type of energy quantity, it will not be used here.) Finally, these gaseous ions are simultaneously transferred into the solution where they interact, in various ways, with the solvent dipoles.

In fact, there exists an electric potential difference between the bulk of the gaseous phase and of the solution. The ions then exert electric work during their

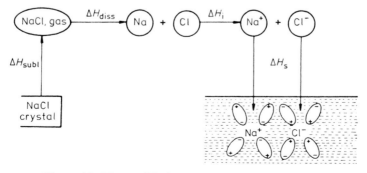

Figure 16. The modified Born–Haber solvation scheme

transfer. However, as cations and anions bearing identical charges of opposite sign are transferred, the electric work for each of them is the same but with opposite sign. Therefore, the resulting total work during simultaneous anion and cation transfer across the electric potential difference between the two media is zero.

This process is generally called *ion solvation* (for solvation in water the term *hydration* is used). Solvation is a process connected with the release of energy (an exothermic process) and therefore the solvation energy ΔH_s has a negative sign. The total energy change during dissolution of the sodium chloride crystal ΔH is given by the sum

$$\Delta H = \Delta H_{subl} + \Delta H_{diss} + I_{Na} + \Delta H_{e, Cl} + \Delta H_s$$

All the quantities on the right-hand side are positive (the corresponding processes are endothermic, i.e. they are connected with the acceptance of energy) except the solvation energy ΔH_s. The sum of the positive quantities is larger than the absolute value of the solvation energy ΔH_s. Therefore, the resulting energy change is positive so that, during the dissolution of sodium chloride, heat must be supplied from an external source or the temperature of the system drops (the internal energy of the system is consumed in the dissolution process).

In the dissolution in water and in methanol all the partial steps are the same, except solvation. The larger decrease in energy (the larger consumption of energy for the dissolution of sodium chloride) in methanol is a result of weaker solvation in this medium (i.e. lower absolute value of the solvation energy ΔH_s).

The existence and extent of ion solvation is reflected in several other observations. For example, ions often move more slowly in solution than would be expected on the basis of their ionic radii determined by crystallographic measurements (cf. Table 1). The velocity of a spherical particle pulled by the force f in a medium of viscosity η follows the Stokes law:

$$v = \frac{f}{6\pi\eta r}$$

where r is the radius of the particle. Assuming that the ion is a spherical particle, we can (very roughly) determine its radius from its mobility, which is evaluated on the basis of conductivity data as shown on page 68.

The difference between the radii determined from crystallographic and mobility data for small ions, such as lithium, is striking. Obviously, such ions have to 'travel' with a very large sheath of water molecules. When hydrated, small ions such as Li^+ have a large resulting size, similar to those with a large charge, e.g. Al^{3+}. On the contrary, the greater the number of electrons in the electron sheath of an ion the weaker its hydration.

Another approach to the study of solvation is the measurement of the thermal capacity of the solution. This quantity depends, among other factors, on the ability of solvent molecules to move from one site to another (translational motion) and to vibrate and rotate. Under the influence of ions present in the solution, these abilities vary, so that the region of action of the ions on the solvent can be assessed by determining the heat capacity of the solution and by comparing it with the capacity of the pure solvent. On the basis of microwave spectra (for example, nuclear magnetic resonance, n.m.r.) it has been found that, after dissolution of an electrolyte in water, some of the water molecules have a different frequency of vibration of the O—H bond than in pure water. These molecules obviously belong to the hydration sheath. It is not surprising that each of the three methods described, i.e. the study of ion mobility, of the heat capacity of the solution and of the microwave spectra of the solvent, gives different information on the extent of solvation, which is sometimes defined by the solvation number (the number of solvent molecules directly effected by the individual ion), as in each method the influence of the ion on a different physical process is measured.

In order to comprehend the process of solvation the properties of the solvent alone will first be considered. In the liquid state, molecules are a very small distance apart and are held together by forces like *dipole–dipole interaction* and *van der Waals attraction*. The van der Waals force is analogous to the interaction between the dipoles but acts between non-polar molecules. Because of vibration, even in a non-polar symmetrical molecule an asymmetrical charge distribution is statistically formed, thus inducing a polar distribution in neighbouring molecules, so that a kind of instantaneous dipole–dipole interaction results (Figure 17). While the force of the dipole–dipole interaction is proportional to r^{-3} (r is the distance between the polar molecules) the van der Waals attraction is proportional to r^{-7}, a typical short-range interaction.

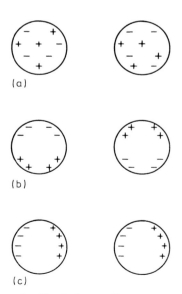

Figure 17. The origin of van der Waals forces. On average, non-polar molecules have a charge distribution shown in (a), but at any arbitrary instant this distribution is asymmetrical so that an instantaneous dipole is formed as in (b) and (c)

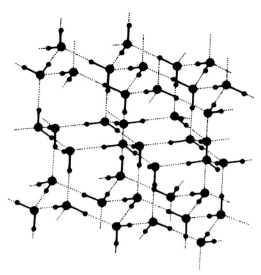

Figure 18. The structure of ice according to L. Pauling. The large circles are oxygen atoms and the smaller ones are hydrogen atoms

Pure van der Waals interactions are present in very simple liquids like liquefied inert gases (neon, argon, etc.) which are, unfortunately, of little interest for fields of chemistry and physics other than for the theory of the liquid state. In liquids that are often used as solvents, other forces simultaneously play an important role. Thus, the rather complicated properties of water may be explained by interactions between dipoles of water molecules and by association of these molecules through hydrogen bonds (hydrogen bridges). As a result of this association clusters of water molecules are formed in liquid water. These clusters, which are reminiscent of the hexagonal structure of ice (see Figure 18), contain different numbers of water molecules (in warmer water smaller clusters prevail). Real disordered 'liquid' water may be between these clusters. The clusters rapidly decompose, reform and are formed again from disordered water, but if this change could be stopped for a moment some kind of water crystals would be observed, as well as irregularly oriented molecules.

If an ion appears in the solvent, its charge and its structure influence the solvent structure in its surroundings. For example, when lithium chloride is dissolved in water the small lithium cation causes the water dipoles to orient with their negative (oxygen) end towards the centre of the ion.* In this way a layer of oriented water molecules is formed in the immediate neighbourhood of the cation, called the *primary hydration sphere* (see Figure 19). Those water molecules align others in the surroundings and a cluster of water molecules is

---

* The angle between O—H bonds in water is 112°, the positive pole of the dipole being situated between the hydrogen atoms while the oxygen atom bears the negative pole.

28

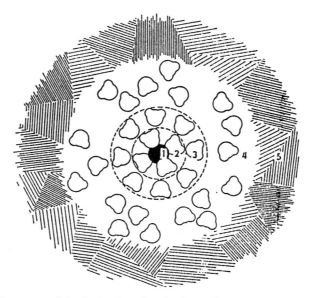

Figure 19. Structure of the hydration sheath of a cation (1). Water molecules immediately in the neighbourhood of the ion are oriented and form the primary hydration sphere (2). In the second hydration sphere the water molecules are slightly aligned with the electric field of the cation (3). The influence of these structures can disorganize the water structure at larger distances (4). At still more distant sites (5) water will behave in the same way as in the absence of ions

formed around the cation, like the instantaneous crystal structure in pure water. There are, however, some differences because the molecules in the primary hydration sphere remain in place for a longer time than water molecules in a cluster that is not influenced by an ion charge. Thus, the cluster around the lithium cation is more stable than the usual cluster in water alone. Since the lithium ion reinforces the water structure, it is termed a *structure-making ion*. When in an aqueous solution the lithium ion is replaced by another alkali metal ion that has a large radius, such as the caesium ion, the result is different. Since this ion has a large electron sheath, the positive charge of the ion acting on the negative ends of the water dipoles is screened by the large electron envelope. At the same time, the electrons repulse the approaching negative ends of the dipoles so that the final result is a disordered distribution of water molecules similar to that in the disarrayed 'fluid' region of water. This property of caesium is termed *structure-breaking*.

Among inorganic ions, chlorides and fluorides are structure-making and nitrates and perchlorates are structure-breaking. Ions that have no pronounced properties of one or the other kind are, for example, the potassium, bromide and sulphate ions.

As shown above, solvation is obviously based on the interaction between the electric charge of the ion and the dipole of the solvent molecule. Substances with larger dipoles exhibit large solvation effects. Since more polar substances have a larger dielectric constant, they will solvate ions more strongly. This has already been made obvious from the comparison of solvation by water and methanol: water, with a dielectric constant about twice as large as methanol, exhibits stronger solvation.

Solvation is not observed only with ions. Polar groups in the molecule (in organic molecules, for example $-C=O$, $-NH$, $-NO_2$, etc.) can also be solvated as a result of dipole–dipole interaction of these groups with the solvent. The solvation of polar groups is important for chemical reactions where these groups participate and, again, decreases with decreasing polarity of the solvent.

When a substance with a typical non-polar group like a long-chain alkyl is dissolved in water, a peculiar phenomenon termed *hydrophobic interaction* is often observed. Such a non-polar group is structure-making and produces an increase in the order of the system. In this way the entropy of the system would (under certain conditions, of course) decrease. Because natural processes tend to increasing disorder (i.e. to an increase in entropy), the long alkyl groups preferably join by van der Waals forces forming complicated associates instead of contacting and ordering water molecules to an ice-like structure. Because of this tendency to escape from the aqueous surroundings these groups were denoted by the term *hydrophobic* (hostile to water). For a quantitative assessment of hydrophobicity, and also of the opposite property, *hydrophilicity*, see page 151. Another closely related phenomenon is concerned with the effect of ions on less polar species which they expel from their solvation sheaths (the *salting-out effect*, resulting in a decrease of solubility of non-polar substances). Because of these two effects the non-polar molecules and, particularly, molecules containing a polar or ionized group in addition to the non-polar part (*amphiphilic* or amphipathic substances; Greek *amphi* = both) preferably enter the surface of solutions (therefore they are called *surfactants*) forming *monolayers*, which is typical of soaps, aqueous solutions of sodium or potassium salts of fatty acids, such as palmitic or stearic acid, as shown in Figure 20(a). At higher concentrations, they form spherical associates called *micelles* (Figure 20b). If a molecule contains one polar and two long-chain alkyl groups, such as various phospholipids, the hydrophobic interaction is so strong that, under conditions stated on page 157, a *lipid bilayer* can be formed.

Complicated solvation effects appear with biopolymers, e.g. proteins. In the solid state proteins form helices, ordered or disordered coils and other structures (see page 53). This arrangement is called the secondary structure of a protein (cf. page 56). The stability of the helix is made possible by hydrogen bonds between carbonyl groups $C=O$ and imide groups $N-H$. On dissolving the protein in water, this situation may change. Both these groups are hydrated, which decreases the probability of hydrogen bond formation, and hydrophobic interactions between the non-polar parts of the molecule prevail, resulting in a transition to a random coil. Conclusions about the secondary structure of a protein obtained, e.g. by solid-state X-ray analysis, cannot be directly applied to the situation existing in solution.

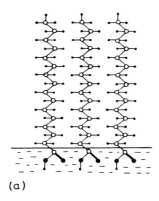

(a)

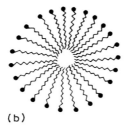

(b)

Figure 20. Monolayers and micelles of amphiphilic molecules. At low concentrations, amphiphilic palmitic acid forms a monolayer (a) at the water/air interface. At higher concentrations spherical structures, micelles (b), are formed

References

1. A. Ben-Naim, *Hydrophobic Interaction*, Plenum Press, New York, 1980.
2. K. Burger, *Solvation, Ionic and Complex Formation in Non-aqueous Solvents*, Elsevier, Amsterdam, 1983.
3. B. E. Conway, *Ionic Hydration in Chemistry and Biophysics*, Elsevier, Amsterdam, 1981.
4. R. R. Dogonadze, E. Kálmán, A. A. Kornyshev and J. Ulstrup (eds.), *The Chemical Physics of Solvation*, Parts A to C, Elsevier, Amsterdam, 1985–1988.
5. I. M. Klotz, 'Structure of water', in *Membranes and Ion Transport* (ed. E. E. Bittar), Vol. I, John Wiley & Sons, New York, 1970.
6. Y. Marcus, *Ion Solvation*, John Wiley & Sons, Chichester, 1986.
7. K. L. Mittal (ed.), *Surfactants in Solution*, Vols. 1 to 3, Plenum Press, New York, 1984.
8. M. J. Rosen, *Surfactants and Interfacial Phenomena*, John Wiley & Sons, Chichester, 1989.
9. 'Solvation', *Faraday Disc. Chem. Soc.*, **85** (1988).
10. C. Tanford, *The Hydrophobic Effect: Formation of Micelles and Biological Membranes*, 2nd ed., John Wiley & Sons, New York, 1980.

NON-ELECTROLYTE SOLUTIONS

In this section we consider solutions from another point of view. The properties of a solution depend strongly on the quantity of the substance dissolved. The representation of this quantity is expressed by various relationships between the amount of dissolved substance (the solute) and the amount of solution or solvent. Quantitative measures of these relationships are various *concentration scales*. Very often the representation of a solute is determined by the amount (in moles) of substance dissolved in a unit volume (usually cubic decimetres or litres) of the solution. In this way, the *molar concentration* is defined in units of mol dm^{-3} (mol l^{-1}). When a solution has the concentration m moles per cubic decimetre we call it m-molar. The ratio of the amount of solute (in moles) to the total amount of all substances in the solution (in moles) is the *molar fraction* (a dimensionless quantity).

If a non-electrolyte (a substance which on dissolution does not dissociate into ions), e.g. cane sugar or sucrose, is dissolved in water and the solution is placed in an evacuated flask, then the gaseous phase above the solution contains only water vapour (at room temperature the evaporation of sucrose can be completely neglected). When the pressure of the water vapour is measured with a manometer it is found that the more the concentration of sucrose increases the lower is the vapour pressure. When the concentration of sucrose is less than about 0.1 mol l^{-1} (the molar fraction is less than 0.001 81), then the pressure of water vapour is decreased by a quantity Δp in comparison with the original pressure of pure water; Δp is directly proportional to the molar fraction of sucrose in the solution. This relationship is called Raoult's law and the solutions obeying it are termed ideal.

The vapour pressure of a liquid depends on the number of molecules in the liquid that reach the surface of the liquid in a certain time interval and then have a chance to evaporate. Sucrose molecules reaching the surface cannot evaporate because the vapour pressure of sucrose is immeasurably low. Thus, sucrose molecules coming to the surface only decrease the probability of evaporation of water molecules, resulting in a decrease in the water vapour pressure.

A similar experiment is illustrated in Figure 21. A vessel has two compartments. Compartment 1 contains only a pure solvent, while a solution of a non-electrolyte with molar concentration c is in compartment 2. The vertical tube T connected to compartment 2 is thin, so that its volume can be neglected with respect to the volume of the solution in 2. Both 1 and 2 are connected by a porous semi-permeable membrane which enables only the solvent molecules to move from one compartment to the other and prevents the solute molecules from passing from 2 to 1. (There are also other types of semi-permeable membranes, as described in Chapter 3, page 139.) When compartment 2 is filled with the solution, the liquid penetrates from 1 to 2 as long as the level of the liquid in tube T reaches a certain height. Thus, the hydrostatic pressure has increased in 2, thereby preventing further penetration of the solvent from 1.

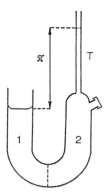

Figure 21. A simple instrument for the study of osmotic pressure (π). Compartment 1 contains pure solvent, compartment 2 the solution

When we increase the solute concentration in 2, the final height of the solvent column in T increases in direct proportion to the concentration.

The chance for penetrating into the pores of the membrane is greater for the molecules of the pure solvent than for the solvent molecules from the solution. In proportion to their number, the solute molecules hinder the solvent molecules from penetrating into the pores of the membrane so that there is a net flow of the solvent molecules from 1 to 2. This phenomenon is called *osmosis*. When the hydrostatic pressure on the solution side is increased, the rate of transfer of the solvent molecules from 2 to 1 increases and counterbalances the flow in the opposite direction. This phenomenon evokes the impression that some kind of negative pressure existing in the solution draws the solvent molecules from the pure solvent. In equilibrium this *osmotic pressure* is counterbalanced by the excess hydrostatic pressure produced.

In dilute solutions the osmotic pressure π is described by the equation

$$\pi = RTc$$

where R is the gas constant, T absolute temperature and c the solute concentration. This equation is identical to the ideal gas law.* Thus the molecules present in a dilute solution behave in an analogous way to the molecules of an ideal gas.

Osmotic pressure was discovered by the botanist W. Pfeffer in 1887. It is of enormous importance for cellular processes. The distribution of water in an organism depends on osmotic pressure equilibrium in the cells and in the intercellular liquid. If a cell is in a medium of lower osmotic pressure than that inside the cell, water present in the intercellular liquid rapidly penetrates into the

---

* Any elementary course in physics or chemistry treats the ideal gas (where no other interactions between the molecules occur besides mutual elastic collisions) by the equation $pV = nRT$, where p is the gas pressure and V the volume of the gas present in an amount of n moles. Since the ratio of n to V gives the concentration c we have $p = RTc$.

cell, thus increasing its volume. In an extreme case the cell will then burst (*plasmoptysis*). On the contrary, if the cell is in a medium of high osmotic pressure than that within the cell, water flows out of the cell, which then shrinks. This phenomenon is called *plasmolysis* (see Figure 22). The changes in the osmotic pressure make possible, for example, the mechanical action of tissues against their surroundings. Thus, the pressure exerted by a growing plant on the neighbouring soil is of this origin. All these phenomena depend on the specific properties of cell membranes as described in Chapter 3.

Returning to the Raoult law, consider a solution with a very high concentration of sucrose. In this range of concentrations the vapour pressure increases more slowly with increasing concentration than would correspond to the ideal behaviour described on page 31, as if the sucrose had a lower concentration than that actually present or as if it were less active. In fact, at higher concentrations of sucrose its molecules interact through dipole–dipole and van der Waals forces, thereby decreasing their influence on water molecules. Solutions whose behaviour is no longer governed by the Raoult law are called *non-ideal*.

Non-ideal properties are also exhibited by the osmotic pressure at higher solute concentrations, which is then lower than would correspond to the above equation. Consequently, with respect to the osmotic pressure the solution behaves as more dilute than it actually is.

At the beginning of this century G. N. Lewis (1875–1946) suggested that, in equations describing various equilibrium properties of solutions like the Raoult law, the osmotic pressure equation and other relationships that will be considered below, concentration quantities like the molar concentration, molar

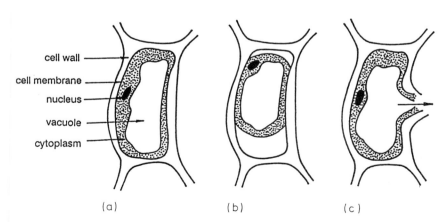

(a)   (b)   (c)

cell wall
cell membrane
nucleus
vacuole
cytoplasm

Figure 22. Plasmolysis. (a) Under normal conditions the cytoplasm spreads to the cell wall which is in contact with the total surface of the cell plasma membrane. (b) In a hypertonic solution (with an osmotic pressure larger than that inside the cell) water flows out of the vacuole and from the cytoplasm and the cell membrane separates from the cell wall. (c) This illustrates the condition in a strongly hypotonic solution from which water is transported into the cell which finally bursts (plasmoptysis)

fraction, etc., should be replaced with *activities*. Activity a is given by the simple relationship

$$a = \gamma c$$

where γ is the activity coefficient and c the concentration. The activity coefficient is a correction factor for non-ideal behaviour.* In dilute solutions it approaches unity and the activity can be identified with concentration. On introducing activities, an equilibrium relationship of the Raoult law type can be used for arbitrary concentrations.

STRONG ELECTROLYTES

The situation is quite different when an electrolyte, e.g. sodium chloride, is dissolved in water. Again the water vapour pressure drops, but the decrease is larger than in the case of a non-electrolyte. For sodium chloride present at a millimolar concentration or lower, the decrease of vapour pressure would correspond to a non-electrolyte solution twice as concentrated. Obviously, on dissolution the sodium chloride completely decomposed or, better, dissociated to chloride and sodium ions. A substance that completely dissociates to ions in a particular solvent is termed a *strong electrolyte*. When the concentration of a strong electrolyte in water is increased over the concentration limit of 10^{-3} mol l^{-1}, the decrease in the water vapour pressure again becomes slower since the ions are less active in the solution than would correspond to their concentration. Why does this happen at considerably lower concentrations than in non-electrolyte solutions? The answer can be found in the electrostatic forces operating between the ions, which act at much larger distances than short-range van der Waals forces. The influence of non-ideal behaviour is again expressed by the activity coefficients. Direct measurement does not yield the activity coefficients of the individual kinds of ions because in solution anions and cations are always simultaneously present in order to preserve the electroneutrality condition. Only the activity coefficient of the whole electrolyte, which is termed the mean activity coefficient $\gamma_{\pm}$, can be found. This limitation need not deter us from attempting to find the activities of individual ions, which are useful for various practical tasks.

The activity coefficients are rather important quantities for they make it possible to describe equilibria in chemical reactions in a fully quantitative way. They appear in expressions characterizing the acidity of solutions, and knowledge of activity coefficients is necessary for the analysis of solutions by potentiometry. Their values for ions present in nerve cells, for example, are important for describing the behaviour of these cells, etc. Consequently, these seemingly rather abstract quantities will be considered in some detail.

---

* It should be noted that the activity coefficients have different values if, for example, a molar fraction scale is used instead of a molar concentration scale.

The electrostatic interaction between ions is stronger the higher the ionic charge. When a solution contains, at the same concentration, types of ions with a higher charge, they will exert a greater electrostatic effectiveness, 'strength', on the components of the solution. This influence was quantitatively expressed by G. N. Lewis as the *ionic strength* I, given by the equation

$$I = \tfrac{1}{2}(z_1^2 c_1 + z_2^2 c_2^2 + z_3^2 c_3 + \cdots)$$

where z_1, z_2, $z_3, \ldots$ are the charge numbers of the individual types of ions present at concentrations c_1, c_2, $c_3, \ldots$. For rather low electrolyte concentrations, Lewis found the empirical relationship

$$- \log \gamma_{\pm} \sim \sqrt{I}$$

The electrolyte solution is only outwardly electroneutral. Within the solution the positively and negatively charged ions move randomly in all directions depending on their thermal energy. At the same time, ions with opposite charges are attracted and those with the same charge are repulsed. Consequently, it can be expected that anions will be found in the neighbourhood of cations, and vice versa. In this way a slight surplus of charge of opposite sign, a space charge, appears in the region around every ion in the electrolyte solution. P. Debye and E. Hückel termed this domain of space charge the *ionic atmosphere*, which is not a static structure but a statistical phenomenon. There is a somewhat greater probability for an ion of opposite charge to appear for a moment in the neighbourhood of the 'central' ion than for an ion with a charge of the same sign. The region of a deviation from electroneutrality extends to only very small distances from the ion (see Figure 23). The thickness of this region is characterized by the Debye length

$$L_{\mathrm{D}} = \sqrt{\frac{D \varepsilon_0 R T}{2 F I}}$$

which is important not only for understanding the processes in electrolyte solutions but also those occurring in colloidal solution, in the electric double layer and at membranes. We have already familiarized ourselves with the quantities D, ε_0, R, T and I. The Faraday constant $F(= 9.65 \times 10^5 \text{ C mol}^{-1})$ is the charge of 1 mol of positive univalent ions.

The ionic atmosphere acts electrostatically on the central ion, thus decreasing its activity. This electrostatic influence increases when the ionic atmosphere becomes more compact. Consequently, in more concentrated solutions the activity coefficient decreases (this is not true for some very concentrated solutions). For the dependence of the activity coefficient on the ionic strength in rather dilute solutions, Debye and Hückel deduced, using a statistical procedure, the formula for an ion carrying the charge z_k

$$\log \gamma_k = - A z_k^2 I^{1/2} \tag{1}$$

and for the mean activity coefficient of an electrolyte (consisting of cations with

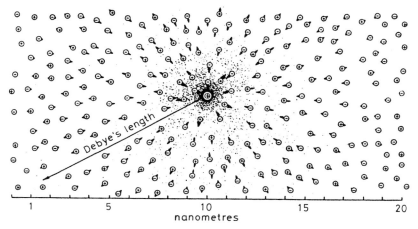

Figure 23. The ionic atmosphere around a cation (in a solution of a univalent electrolyte with $c = 10^{-3}$ mol dm^{-3}). Arrows of different size indicate electrostatic interaction with the cation. The shaded area indicates the presence of space charge

the charge z_+ and anions with the charge z_-)

$$\log \gamma_\pm = z_+ z_- A I^{1/2} \tag{2}$$

Equation (1) is valid for the activity coefficient of an individual ion because it was obtained using a non-thermodynamic approach. The mean activity coefficient of a uni-univalent electrolyte is given by the relationship

$$\gamma_\pm = \sqrt{\gamma_+ \gamma_-}$$

where γ_+ and γ_- are the activity coefficients of the cation and anion, respectively. A is a function of temperature and of several basic physical constants; at 25 °C, $A = 0.51$ l$^{1/2}$mol$^{1/2}$. The actual activity coefficient deviates at rather low values of the ionic strength from the dependence described by equations (1) and (2). These simple relationships, usually called the Debye–Hückel limiting law, correspond to a grossly simplified picture of a solution of a strong electrolyte. The ions are represented by point charges (the influence of the size of the ion is neglected)—each of them can approach the other to an arbitrary distance and exclusively electrostatic forces act among them. It is necessary, however, to stress that the Debye–Hückel limiting law exactly describes a simple model of a natural situation.

When the electrolyte becomes more concentrated, the ions interact in various ways. They are, of course, not point charges but have characteristic dimensions. An individual ion can only penetrate to the neighbourhood of another ion to a distance given by the sum of their radii. The ions in solution are not completely 'free' but are solvated. The interaction of solvated ions is not exclusively electrostatic but, particularly at higher concentration, short-distance forces play a role. At still higher concentrations, the ions compete for solvent molecules. In various solvents *ion association* occurs to a different extent. As shown by N. Bjerrum, under definite circumstances two ions of opposite charge can exist more advantageously as an electroneutral aggregate bound together by electrostatic forces than as two separate ions in solution. Such an ion-pair is formed more easily at higher concentrations of the electrolyte and at lower permittivity of the solvent. In water ($D = 78$) ion-pairs appear at high (at least one molar) electrolyte concentrations. However, an exact concentration limit for ion-pair formation in water cannot be determined because the effects of ion interaction are superimposed on the influence of the association. Thus, there are no adequate means of separating the individual influences. In solvents with a lower permittivity, such as nitrobenzene, acetonitrile or methanol (D around 30), ion association is marked at concentrations of the order of 10^{-2} mol dm^{-3}. Finally, in dioxane ($D = 4$) or in benzene ($D = 2$) the ions are almost completely associated at quite low concentrations. In addition to simple anion–cation associates, even more complicated structures like two cations–one anion, etc., can be present in these solvents.

Thus, for example, 3×10^{-3} M tetraisopentylammonium nitrate in dioxane is only dissociated by about 0.000 003%. In the medium of the inside of a biological membrane, formed predominantly by long-chain alkyl groups (see page 157), all ions are practically associated.

Ion association also depends, of course, on factors other than mere interaction between ion changes. A definite role is played by ion–dipole and van der Waals forces, both of which depend on the structure of the ions.

It is difficult to develop a complete theory of strong electrolytes that would consider all the factors mentioned. The present approach, based on the methods of statistical mechanics, first considers pair interactions (i.e. the values of the energy necessary for the approach of two ions at varying concentration and solution conditions).

Equation (1) is a more general expression of the activity coefficient of an individual ion than is obvious from the preceding considerations. Let us determine, for example, the activity coefficient of the zinc ion in a solution of 10^{-5} M $Zn(NO_3)_2$ and 10^{-3} M KNO_3. (The symbol M KNO_3 will denote a 'molar solution of KNO_3'.) The ions derived from the potassium nitrate do not react with the zinc ion and only exhibit an influence on its activity coefficient which depends on the total ionic strength alone. Such an excess electrolyte is termed *indifferent*. For the calculation of the activity coefficient of the zinc ion, equation (1) can then be applied directly.

When dissolved in a liquid a *weak electrolyte* is only partially dissociated into ions. The degree of dissociation depends on the equilibrium constant, as we shall learn in more detail in the section on acids and bases.

References

1. Page xiii, Ref. 2, Chapter 1.
2. R. W. Guerney, *Ionic Processes in Solution*, McGraw-Hill Book Company, New York, 1953.

3. H. S. Harned and B. B. Owen, *The Physical Chemistry of Electrolytic Solutions*, Reinhold, New York, 1967.
4. A. L. Horvath, *Handbook of Aqueous Electrolyte Solutions, Physical Properties, Estimation and Correlation Methods*, Ellis Horwood, Chichester, 1985.
5. V. M. M. Lobo, *Handbook of Electrolyte Solutions*, Parts A and B, Elsevier, Amsterdam, 1990.
6. R. A. Robinson and R. H. Stokes, *Electrolyte Solutions*, Butterworths, London, 1959.

ACIDITY OF SOLUTIONS

When eating a sour cherry we experience an acid taste. The taste of 0.1% hydrochloric acid (a more concentrated acid would cause burns) is stronger (approximately the same as that of stomach juices, which contain about 0.1% hydrochloric acid). A glycine (aminoacetic acid) (Greek *glykys* = sweet) solution has a sweetish taste although it is termed an acid. Estimation by tasting, though very accurate with experienced people, cannot be used as an objective measure of acidity. The appropriate methods of quantitative assessment of acidity will now be considered.

Already in the distant past a link was suspected between the acidity and the composition of acid substances. At the end of the eighteenth century the French chemist Lavoisier suggested that the acid substance was oxygen. (Lavoisier formed the name of oxygen from two Greek words, *oxys* = acid, sharp, and *gennao* = bear, produce). However, as early as the beginning of the nineteenth century the English scientist Humphry Davy showed that the very acidic hydrogen chloride only consists of chlorine and hydrogen and that hydrogen is the substance common to all acids. Finally, at the end of the nineteenth century, the Swedish chemist Svante Arrhenius defined acids as compounds that can split off hydrogen ions.

A common school demonstration is the neutralization of a solution of hydrochloric acid with sodium hydroxide as a base (i.e. the transformation of an acid into a salt). An indicator (e.g. bromothymol blue) is added to the solution. By a colour change (in this case from yellow to blue) this substance indicates the equivalence point where an equal concentration of the acid and the base is present in the solution. Let us carry out this experiment in a different way. After

each addition of the solution of sodium hydroxide the conductance of the solution will be measured. In this way the dependence shown in Figure 24 is obtained. The conductance of a solution is roughly proportional to the number of ions present and to their mobilities. The added solution of sodium hydroxide is very concentrated, so that the additions result in only a negligible change in the total volume of the solution. Since hydrochloric acid is a strong electrolyte, neutralization results in the loss of a strongly conductive component from the solution, which is replaced by much less conductive ions. As there are further supporting facts for the view that the hydrogen ion is the most mobile of all ions, the neutralization process can be identified with the reaction

$$H^+ + OH^- = H_2O$$

Thus, in the solution the hydrogen ions are replaced by much less mobile sodium ions coming from sodium hydroxide and the conductance decreases. The increase of conductance following the equivalence point is caused by sodium and hydroxide ions from excess sodium hydroxide.

Another situation occurs with neutralization of acetic acid (Figure 24). The initial increase of conductance corresponds to the neutralization of the hydrogen ions formed by *dissociation* of the acid in the original solution. There are, however, only a few hydrogen ions in the solution and, when they have been consumed, the conductance starts to increase again as the undissociated acid reacts with the hydroxide ions:

$$CH_3COOH + OH^- = CH_3COO^- + H_2O$$

Acetic acid differs from hydrochloric acid in the simultaneous presence of undissociated CH_3COOH molecules together with acetate and hydrogen ions. Hydrochloric acid, which is completely split into hydrogen and chloride ions, is called a strong acid. In contrast, acetic acid, which is, in part, present in the

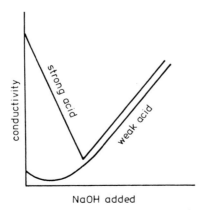

Figure 24. Dependence of the conductance of a solution on the degree of neutralization of a strong and a weak acid (conductometric titration curve)

solution in an undissociated form, is a weak acid. It will be shown below that the strength of acids depends on the medium, i.e. on the solvent.

The hydrogen ion cannot be considered as an isolated proton since it has a structure that depends on the solvent in which the acid dissociates. In water this structure is H_3O^+, the oxonium (hydronium) ion; in pure 'glacial' acetic acid it is $CH_3COOH_2^+$; in concentrated sulphuric acid, $H_3SO_4^+$; etc. The basic rules characterizing the acidity of solutions were formulated in 1913 by the Danish chemist J. N. Brønsted. The acid as well as the basic properties become apparent when the substance reacts with the solvent and depends on it. Substances that split off protons in a solution are acids. Thus, for example, hydrogen chloride or acetic acid react with water according to the equations

$$HCl + H_2O = H_3O^+ + Cl^-$$

$$CH_3COOH + H_2O = H_3O^+ + CH_3COO^-$$

By an analogous experiment to that with acetic acid, we find that ammonia behaves as a base in aqueous solutions and accepts a proton from the solvent:

$$NH_3 + H_2O = NH_4^+ + OH^-$$

However, in concentrated sulphuric acid, acetic acid acts as a base, i.e.

$$CH_3COOH + H_2SO_4 = CH_3COOH_2^+ + HSO_4^-$$

In water, aniline is a base:

$$C_6H_5NH_2 + H_2O = C_6H_5NH_3^+ + OH^-$$

while in liquid ammonia it behaves as an acid:

$$C_6H_5NH_2 + NH_3 = C_6H_5NH^- + NH_4^+$$

If the reaction of an acid HA with a solvent HS proceeds in a not very dilute solution in such a way that almost all of the dissolved substance dissociates or, in other words, the equilibrium

$$HA + HS = A^- + H_2S^+ \tag{1}$$

is considerably shifted to the right-hand side, this substance is termed a *strong acid*. If, on the contrary, the resulting concentrations of the reaction products A^- and H_2S^+ are much lower than that of the original substance HA, we call it a *weak acid*. In an analogous way a base reacts with the solvent according to

$$B + HS = BH^+ + S^- \tag{2}$$

With respect to the extent to which this equilibrium is shifted to the right-hand side, the base is termed strong or weak. Obviously in the case of a base no dissociation takes place but a transformation from an uncharged form B to the charged form BH^+ occurs. Therefore, it is more appropriate to use the term *ionization* for acid–base equilibria.

Evidently, the anion of an acid acts as a base according to the equation

$$A^- + HS = HA + S^-$$

Therefore the anion A^- is termed the conjugate base to the acid HA, in the same way as BH^+ is the conjugate acid to the base B. The *protolytic reaction* is the reaction of an acid with a solvent or with another basic species to form a conjugate base. The acids can also be denoted as proton donors and the bases as proton acceptors.

Acids that split off several protons (oligoprotonic acids, Greek *oligoi* = several), such as, for example, orthophosphoric acid H_3PO_4, undergo protolytic reactions to give species that can act as acids or bases at the same time. Thus, the $H_2PO_4^-$ anion reacts in water as an acid:

$$H_2PO_4^- + H_2O = HPO_4^{2-} + H_3O^+$$

or as a base:

$$H_2PO_4^- + H_2O = H_3PO_4 + OH^-$$

Reference

1. Page xiii, Ref. 2, Section 1.4.

PROTONS IN ACTION—DISSOCIATION RATES AND EQUILIBRIA

Let us consider the equilibrium of the protolytic reaction (1) in the last section in more detail. First, the rate of each of the two reactions will be discussed; in other words, the kinetic aspect of the equilibrium will be taken into account. Proceeding from the left to the right reaction (1) is called the dissociation reaction and the opposite reaction is the recombination of the acid anion with the solvated proton. Thus, the rate of production of the solvated proton can be considered as a measure of the protolytic reaction, i.e.

Rate of protolysis (total increment of concentration of solvated protons per second) = rate of dissociation − rate of recombination

According to the Guldberg–Waage law (or the mass action law), the reaction rate is given by the product of the concentrations of the reacting species (or reactants) multiplied by a rate constant. Thus,

Rate of dissociation

= k_d [HA][HS] (increment of concentration of solvated protons per second)

Rate of recombination

= k_r [A$^-$][H$_2$S$^+$] (decrease of concentration of solvated protons per second)

Symbols in brackets denote concentrations (usually in moles per litre) of the substance indicated. Since both of these reactions are rapid their rate constants,

k_d for dissociation and k_r for recombination, are very high (the recombination rate constant is the largest attainable in a solution). In equilibrium there is no net production of solvated protons; therefore

$$\text{Rate of protolysis} = 0$$

and the rates of both reactions cancel each other (this does not mean that they do not proceed with undiminished velocity):

$$\text{Rate of dissociation} = \text{rate of recombination}$$

$$k_d\,[\text{HA}][\text{HS}] = k_r\,[\text{A}^-][\text{H}_2\text{S}^+]$$

Finally, the equilibrium dissociation constant, K, is given by

$$K = \frac{k_d}{k_r}$$

The solvent HS is always present in great excess and can be included in the final value of the equilibrium dissociation constant, $K_d = K\,[\text{HS}]$, giving

$$K_d = \frac{[\text{A}^-][\text{H}_2\text{S}^+]}{[\text{HA}]}$$

or, in aqueous solution,

$$K_d = \frac{[\text{A}^-][\text{H}_3\text{O}^+]}{[\text{HA}]} \tag{1}$$

For a more exact description of these equilibria, activities should be employed instead of concentrations.

The larger the K_d, the greater is the dissociation and the stronger is the acid. For acids that are strong electrolytes, the dissociation is complete and the equilibrium constant meaningless. This class of acids includes perchloric, chloric, hydrochloric, nitric, sulphuric acids (for the first degree of dissociation $\text{H}_2\text{SO}_4 + \text{H}_2\text{O} = \text{HSO}_4^- + \text{H}_3\text{O}^+$), etc.—of course, in aqueous solutions. The strength of an acid can be advantageously characterized by the logarithm of the dissociation constant multiplied by -1. This quantity is denoted by the symbol

$$-\log K_d = pK_d$$

Obviously, the larger the pK_d, the weaker is the acid. Table 4 lists the pK_d values of more important weak acids. The relations for the ionization constants of bases can be expressed in a similar way using equation (2) in the last section. There is, however, an advantage in using equation (2) in the direction from right to left, and in characterizing the ionization of a base by the dissociation constant of the conjugate acid BH^+, which yields, for example, for ammonia,

$$K_d = \frac{[\text{NH}_3][\text{H}_3\text{O}^+]}{[\text{NH}_4^+]}$$

Table 4. Dissociation constants of weak acids and cations of weak bases (only the name of the base, e.g. ammonia, not ammonium ion, is indicated) according to P. Vaný́sek from *Handbook of Chemistry and Physics* (eds. C. Weast, M. J. Astle and W. H. Beyer), 67th edition, CRC Press, Boca Raton, 1986

Acid	pK_d	°C	Base	pK_d	°C
Formic	3.75	20	Acetamide	0.63	25
Acetic	4.75	25	Aniline	4.63	25
Chloroacetic	2.85	25	α-Naphthylamine	3.92	25
Dichloroacetic	1.48	25	β-Naphthylamine	4.16	25
Trichloroacetic	0.70	25	o-Chloroaniline	2.65	25
Propionic	4.87	25	m-Chloroaniline	3.46	25
Benzoic	4.19	25	p-Chloroaniline	4.15	25
o-Phosphoric	2.12	25	Ammonia	4.75	25
Dihydrogen phosphate,			Methylamine	10.66	25
$H_2PO_4^-$	7.21	25	Dimethylamine	10.73	25
Monohydrogen phosphate,			Trimethylamine	9.81	25
HPO_4^-	12.67	25	Ethylamine	10.81	20
Carbonic[a]	6.37	25	Propylamine	10.71	20
Monohydrogen carbonate	10.25	25	Hydrazine	6.05	25
Boric	9.14	20	Hydroxylamine	8.04	25
Hydrogen sulphide	7.04	18			
Monohydrogen sulphide	11.96	18			

[a] Carbonic acid is, in fact, a stronger acid as its dissociation constant is determined for the overall concentration $c_{H_2CO_3} = [H_2CO_3] + [CO_2]$. Carbon dioxide is only partially transformed to the undissociated carbonic acid (which is an acid of medium strength with $pK = 3.58$); cf. page 47.

The higher the pK_d the stronger is the base. Examples of pK_d values for more important bases are also listed in Table 4.

When the conductivity of water purified by multiple distillation in a quartz or platinum still is measured, we find that, in spite of maximum control of the purity, a definite small conductivity remains. This conductivity is due to the *self-ionization* of water:

$$2\,H_2O = H_3O^+ + OH^-$$

The equilibrium constant defined in the same way as in equation (1) above is called the ion product of water:

$$K_w = K[H_2O]^2 = [OH^-][H_3O^+] = 10^{-14}\ mol^2\ dm^{-6} \tag{2}$$

at 25 °C.

References

1. Page xiii, Ref. 2, Chapter 1.
2. A. Albert and E. P. Sergeant, *The Determination of Ionization Constants. A Laboratory Manual*, 3rd ed., Chapman and Hall, London, 1983.
3. D. D. Perrin, *Dissociation Constants of Inorganic Acids and Bases in Aqueous Solutions*, Butterworths, London, 1969.

PROTONS IN VARIOUS SOLVENTS

In an analogous way, the self-ionization occurs in methanol, acetic acid, liquid ammonia, and many other solvents according to the general equation

$$2\,HS = H_2S^+ + S^- \tag{1}$$

These solvents are called protic since the self-ionization proceeds by proton transfer from one solvent molecule to another. Aprotic solvents contain either no hydrogen atoms at all, like liquid sulphur dioxide, or these atoms are so strongly bound that proton transfer is not feasible. This group includes important solvents like benzene, dioxane

$$O{<}^{\displaystyle CH_3-CH_2}_{\displaystyle CH_3-CH_2}{>}O$$

dimethylsulphoxide $(CH_3)_2SO$, acetonitrile CH_3CN, dimethylformamide $HCON(CH_3)_2$ and hexamethylphosphortriamide $[(CH_3)_2N]_3PO$.

In protic solvents the proton of an acid is bound to a solvent molecule with a different strength. Therefore the equilibrium of equation (1) is shifted to a varying extent to the right-hand side, depending on the nature of the acid and of the solvent. As already mentioned, different substances act as acids of varying strength, or even as bases, depending on the character of the solvent. Water, an amphiprotic solvent, exhibits both acid and basic properties. Solvents like ammonia that are bases in aqueous solution will readily split off protons from the acids and, in this way, increase their strength but, on the contrary, will decrease the strength of the bases (in comparison with water). Solvents of this sort are termed *protophilic*.

On the other hand, solvents that act as acids in aqueous solutions (e.g. acetic acid) tend to weaken acids and strengthen bases. They are called *protogenic* solvents.

As mentioned on page 37, the low permittivity of a solvent encourages ion-pair formation. In such solvents all acids and all bases are weak because full ionization is impossible. Because of their low permittivity, *aprotic* solvents like benzene or dioxane are unable to solvate protons. No protolytic reaction can occur in such media. This is, of course, not true for all aprotic solvents because solvents with high permittivity like dimethylformamide, dimethylsulphoxide or acetonitrile are suitable media for protolytic reactions of acids. They are called *basic aprotic solvents*.

References

1. A. K. Covington and T. Dickinson (eds.), *Physical Chemistry of Organic Solvent Systems*, Plenum Press, New York, 1973.
2. G. Mamantov (ed.), *Characterization of Non-aqueous Solvents*, Plenum Press, New York, 1978.
3. B. Trémillon, *Chemistry in Non-aqueous Solvents*, Reidel, Dordrecht, 1974.

ACTIVITY OF HYDROGEN IONS

The acidity of aqueous solutions is the result of the presence of the oxonium (hydronium) ion, H_3O^+ (it is usual to call it the hydrogen ion). The solvation of protons in water is not completed by formation of H_3O^+ ions because more complex structures are formed, the most stable being $H_9O_4^+$:

$$
\begin{array}{c}
\text{H}\quad\text{H} \\
\diagdown\!\diagup \\
\text{O} \\
\vdots \\
\text{H} \\
| \\
\text{O}^+ \\
\diagup\quad\diagdown \\
\text{H}\quad\quad\text{H} \\
\text{H}\!-\!\text{O}\qquad\quad\text{O}\!-\!\text{H} \\
|\qquad\qquad\quad| \\
\text{H}\qquad\qquad\text{H}
\end{array}
$$

The concentration, or, more accurately, the activity, of oxonium ions decisively determines the course of very many chemical reactions and is of profound importance for processes in living organisms. The hydrogen ion concentration in blood plasma is 4×10^{-8} mol dm^{-3}, while in stomach juices it reaches a value of 0.03 mol dm^{-3}. The growth of some organisms is possible in a very narrow range of oxonium activities, while the reproduction of others (moulds, for example) is only prevented by either extremely high or extremely low oxonium ion concentrations.

In aqueous solutions the interval of the values of oxonium activities is very broad. In 10^{-3} M HCl it approaches a value of 10^{-3} mol dm^{-3} (in fact, it is somewhat lower since the activity coefficient is 0.97). The value of the oxonium activity in 10^{-3} M NaOH can be calculated using the ion product of water (equation (2), page 43). The hydroxide ion activity in this solution is 0.97×10^{-3} mol dm^{-3} and equation (2) yields $a_{H_3O^+} = 1.03 \times 10^{-11}$ mol dm^{-3}. In view of this enormous range of oxonium ion activities it is preferable not to give the activities themselves but their logarithms multiplied by -1.*

This quantity is denoted as pH

$$pH = -\log a_{H_3O^+} \approx -\log[H_3O^+]$$

---

* In textbooks on thermodynamics the activity is usually defined as a dimensionless number, the dimension of the activity coefficient being the reciprocal of the appropriate concentration quantity. In a book destined for non-specialists in physical chemistry, it seems more advantageous to consider the activity as a kind of concentration corrected for various interactions in the system under investigation. The main difficulty in this treatment arises from the logarithmic functions of the activity, because mathematicians require dimensionless quantities as variables in logarithmic functions. We shall circumvent this inconsistency by tacit assumption that in functions like log a the activity a (with units of moles per cubic decimetre, for example) is, in fact, divided by unit activity so that the operator log is applied to a dimensionless quantity.

This simple definition is connected with a certain complication inherent in the fact that $a_{H_3O^+}$ is the activity of an individual kind of ion. However, this question will not be considered in detail, and it will simply be assumed that the activities of oxonium ions exist and that it is completely satisfactory to use the Debye–Hückel limiting law (page 35) for determining the pH of very dilute strong acids and alkali hydroxides. Furthermore, the pH values of certain solutions have been standardized (see Table 5) and the oxonium activities of unknown solutions are determined after calibration of a glass electrode with these standards.

The pH of a 2.7×10^{-4} M HCl solution is 3.73. When sodium hydroxide is added to the solution so that its resulting overall concentration is 3.7×10^{-4} M the pH of the solution reaches a value of 10. The sodium hydroxide has neutralized the oxonium ions coming from the hydrochloric acid and the remaining concentration, 10^{-4} mol dm^{-3} of hydroxide ions, decides the final pH value. In a second experiment, a solution containing 5×10^{-2} M acetic acid and 5×10^{-3} M sodium acetate is prepared, with a pH of 3.73. When the same amount of sodium hydroxide as in the preceding experiment is added to the solution, the pH changes only slightly to 3.77. The mixture of a weak acid and a weak acid salt keeps the pH close to the original value. The same influence is exhibited when hydrochloric acid is added in the same amount. Thus, this

Table 5. pH of five standard buffer solutions at different temperatures. (From R. G. Bates, *Determination of pH. Theory and Practice*, 2nd ed., John Wiley & Sons, New York, 1973)

°C	A	B	C	D	E
0		4.003	6.984	7.534	9.464
5		3.999	6.951	7.500	9.395
10		3.998	6.923	7.472	9.332
15		3.999	6.900	7.448	9.276
20		4.002	6.881	7.429	9.225
25	3.557	4.008	6.865	7.413	9.180
30	3.552	4.015	6.853	7.400	9.139
35	3.549	4.024	6.844	7.389	9.102
40	3.547	4.035	6.838	7.380	9.068
45	3.547	4.047	6.834	7.373	9.038
50	3.549	4.060	6.833	7.367	9.011
55	3.554	4.075	6.834		8.985
60	3.560	4.091	6.836		8.962
70	3.580	4.126	6.845		8.921
80	3.609	4.164	6.859		8.885
90	3.650	4.205	6.877		8.850
95	3.674	4.227	6.886		8.833

A = potassium hydrogen tartarate (sat. at 25 °C).
B = potassium hydrogen phthalate (0.05 mol kg^{-1}).
C = KH$_2$PO$_4$ (0.025 mol kg^{-1}) + Na$_2$HPO$_4$ (0.025 mol kg^{-1}).
D = KH$_2$PO$_4$ (0.008695 mol kg^{-1}) + Na$_2$HPO$_4$ (0.03043 mol kg^{-1}).
E = Na$_2$B$_4$O$_7$ (0.01 mol kg^{-1}).

mixture acts as a kind of *buffer* against the invading hydronium or hydroxide ions. In the solution of 5×10^{-2} M acetic acid alone the acid is dissociated only to a small degree. Again when sodium acetate alone is present in the solution, it completely dissociates to sodium cations and acetate anions. These anions further hydrolyse to a small degree, i.e. they combine with water protons. Some undissociated acetic acid is formed and hydroxide ions are set free:

$$CH_3COO^- + H_2O = CH_3COOH + OH^-$$

Therefore a solution of a weak acid (CH_3COOH) with a strong base (NaOH) is alkalinous. When excess acetic acid is present, it suppresses the hydrolysis and shifts the equilibrium to the left-hand side. On the other hand, the excess acetate ions decrease the dissociation of the acetic acid. In the final result the excess acetic acid is virtually in an undissociated form, while the concentration of the excess salt is equal to the concentration of the free anion of the acid. Thus, for calculation of the pH of the solution, equation (1) (page 42) can be used which, in the logarithmic form, yields the Henderson–Hasselbalch equation:

$$pH = pK_d + \log\left(\frac{[A^-]}{[HA]}\right)$$

$$\approx pK_d + \log \frac{\text{salt concentration}}{\text{weak acid concentration}} \tag{1}$$

This equation indicates the pH of a mixture of a weak acid (acidic component) and its salt with a strong base (basic component) called a buffer. In an analogous way a mixture of a weak base and its salt with a strong acid also acts as a buffer (e.g. ammonia/ammonium chloride).

If a definite amount of a strong base is added to the buffer, part of the undissociated acid is transformed to the salt. The new pH value can be calculated using equation (1). The extent of the pH change is smaller, the smaller the concentration of the newly formed salt compared with the original concentrations of the acid and the salt. The buffering capacity increases with increasing concentration of both components of the buffer. Table 6 contains some recommended buffer solutions.

The preservation of a constant oxonium activity using buffers is particularly unavoidable in the study of organic reactions, in biochemistry and in microbiology. The buffering systems are created by the organism itself. Blood plasma maintains a pH between 7.35 and 7.45. It contains various species of acidic and basic character such as proteins and organic acids, but the main influence is exhibited by the buffer carbonic acid, H_2CO_3/bicarbonate ion, HCO_3^-. The acidic component, H_2CO_3, is the product of oxidation of a variety of carbonaceous substances in the organism. The original product, carbon dioxide, which is in equilibrium with the carbonic acid, escapes from the blood in the lungs. The transformation of carbonic acid to carbon dioxide is a comparatively slow reaction but it is catalysed by the enzyme carbonate dehydratase. Because of the continuous removal of carbon dioxide from blood, the concentration of carbonic acid is considerably lower than that of the bicarbonate anion

Table 6. Buffers with constant ionic strength $I = 0.2$. (From H. M. Rauen (ed.), *Biochemisches Taschenbuch*, Vol. II, Springer-Verlag, Berlin, 1964, p. 104)

pH	Basic solutions[a]								
	1	2	3	4	5	6	7	8	9
2.0	72.0	10.6	14.7						
2.5	72.0	22.8	8.6						
3.0	72.0	31.6	4.2						
3.5	72.0	36.6	1.7						
4.0	72.0				20.0	33.7			
4.5	72.0				20.0	11.5			
5.0	72.0				20.0	3.7			
5.5	72.0				20.0	1.2			
6.0	72.0						9.2	6.6	
6.5	72.0						16.6	3.7	
7.0	72.0						22.7	1.6	
7.5	72.0						24.3	0.5	
8.0	72.0		10.4						80.0
8.5	72.0		5.3						80.0
9.0	72.0		2.0						80.0
9.5	72.0	34.5		2.7					
10.0	72.0	28.8		5.6					
10.5	72.0	23.2		8.4					
11.0	72.0	19.6		10.2					
11.5	72.0	17.6		11.2					
12.0	72.0	15.2		12.4					

[a] The indicated volume of basic solutions (in cubic centimetres) is made up to 200 cm³. Basic solutions: 1, 5 M NaCl; 2, 1 M glycine; 3, 2 M HCl; 4, 2 M NaOH; 5, 2 M sodium acetate: 6, 3.5 M acetic acid; 7, 0.5 M Na_2HPO_4; 8, 4 M NaH_2PO_4; 9, 0.5 M sodium salt of veronal.

(0.027 mol dm⁻³), so that the resulting pH is higher than 7 (the pK of carbonic acid is 6.1).

Aqueous solutions of salts of strong acids and bases (e.g. KCl and Na_2SO_4) should exhibit a neutral pH (pH = 7). However, such salt solutions exposed to the air are always acidic, with a pH of about 5.5. This seemingly surprising finding is due to dissolution of omnipresent carbon dioxide in water, with a resulting acidic reaction.

The meaning of the pH in minute intracellular structures, called organelles, is often questioned (the mitochondrion, chloroplast, endoplasmic reticulum, etc., belong to this group). For the sake of simplicity consider a spherical organelle with a diameter of 0.1 μm. Its volume is approximately 4×10^{-21} m³ $= 4 \times 10^{-18}$ dm³. If the inside of the organelle has a pH of 7, it only contains approximately 4×10^{-25} mol H_3O^+, i.e. only one-quarter of an ion. This result might lead to the conclusion that the pH of such small structures is meaningless.

Let us consider the simple case that a buffer present in the organelle consists of a weak acid (pK = 5), the concentration of the free acid being 10^{-4} mol dm⁻³ and of the acid anion 10^{-3} mol dm⁻³. The dissociation rate constant multiplied by the solvent (water) concentration k_r $[H_2O] = 10^4\,s^{-1}$ and the recombination rate constant $10^9\,dm^3\,mol^{-1}\,s^{-1}$ (cf. page 41). Thus, during 1 second, 1 mole of oxonium ions, i.e. 6×10^{23} individual oxonium ions, are produced in $1\,dm^3$, corresponding to 2.4×10^6 oxonium ions in the organelle. However, their lifetime is short. In view of the reaction rate theory this quantity is $(k_r\,[A^-])^{-1} = 10^{-6}\,s$. This example shows that pH is a statistical quantity depending, in the first place, on the concentrations of proton donors and proton acceptors present in the structure. There is no need for scepticism because of 'fractions of a proton'.

References

1. R. G. Bates, *Determination of pH. Theory, Practice*, John Wiley & Sons, New York, 1973.
2. A. K. Covington, R. G. Bates and R. A. Durst, 'Definition of pH scales, standard reference values, measurement of pH and related terminology', *Pure Appl. Chem.*, **57**, 531 (1985).
3. A. Kotyk and J. Slavík, *Intracellular pH and Its Measurement*, CRC Press, Boca Raton, 1989.

MEASUREMENT OF INTRACELLULAR pH

The method most used for pH determination employs the glass electrode, the properties of which will be discussed on pages 145–146. There is, of course, a number of different methods, including the rather old method of colour indicators, for a long time dismissed but recently reintroduced with great success.

Many organic dyes function as acids or bases. Their protolytic reaction is accompanied by a striking change in colour which is obviously connected with a more radical change in their molecule than would correspond to a mere dissociation or binding of a proton. For example, phenolphthalein does not absorb visible light in the acid and neutral region, but at pH > 8.2 one of its hydroxyl groups becomes ionized and, at the same time, a tautomeric* change takes place:

Colourless form   Red form

---

* Tautomerization (Greek *tautos* = identical, *meros* = part) is a reaction in which the molecules of the reactant and of the reaction product have the same number and kind of atoms but different structure.

The quinoid group C=⟨quinoid ring⟩=O is a chromophore (Greek *chroma* = colour, *ferein* = carry) which absorbs light radiation in the whole visible region of the spectrum with the exception of the long-wavelength range (red light). On absorption by the quinoid system the light energy is dissipated (devaluated) in the form of the energy of thermal movement or solvent molecules. The transition of the undissociated form to the quinoid anion form is characterized by an equilibrium constant which has similar features to the dissociation constant of an acid. The ratio of the concentration of the undissociated (colourless) form and of the quinoid form indicates the pH of the solution. Even here the Henderson–Hasselbalch equation (page 47) can be used.

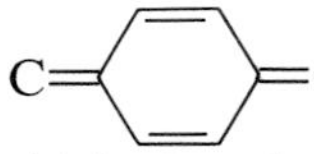

A remarkable success was achieved with an analogous method for determination of intracellular pH according to J. Slavík and A. Kotyk. Fluorescein is a fluorescence indicator, which means that in contrast to the preceding case the energy is not completely degraded but is partially transformed to fluorescence, radiation emission with a lower energy. The basic substance used for the determination of intracellular pH is fluorescein diacetate:

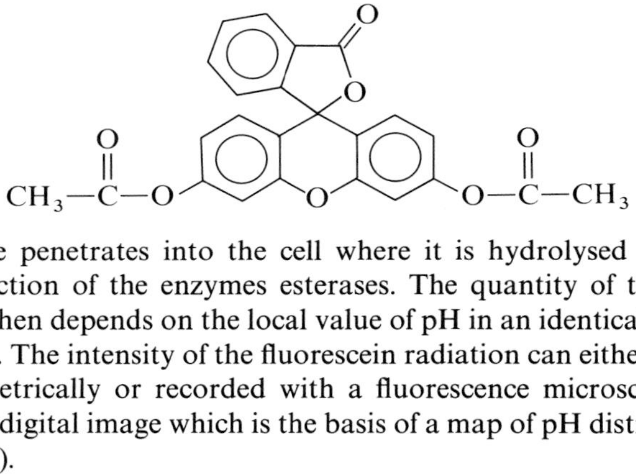

This substance penetrates into the cell where it is hydrolysed to fluorescein through the action of the enzymes esterases. The quantity of the fluorescent quinoid form then depends on the local value of pH in an identical way as in the preceding case. The intensity of the fluorescein radiation can either be measured spectrofluorimetrically or recorded with a fluorescence microscope and then processed to a digital image which is the basis of a map of pH distribution in the cell (Figure 25).

GIANT ACIDS AND BASES

Fatty acids substituted by amine and eventually by other groups, the amino acids, are the building units of proteins and of simpler peptides that often have important biological functions, e.g. hormones and antibiotics. The simplest amino acid is glycine or aminoacetic acid (2-aminoethoic acid). Its formula according to this name would be NH_2CH_2COOH, which obviously has an acidic group ($-COOH$) as well as a basic group ($-NH_2$) in its molecule. Such substances are called *ampholytes*. It is then of interest to find the conditions under which the acidic or the basic function becomes decisive. In a strongly acidic medium glycine is present in the form $NH_3{}^+CH_2COOH$. From the

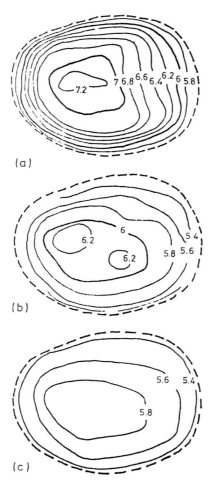

Figure 25. pH maps of *Saccharomyces cerevisiae* cell transferred from water to a 0.2 M buffer of pH 3: (a) after 2 min; (b) after 5 min; (c) after 20 min (Reproduced by permission of J. Slavík and A. Kotyk)

measurement of the ionization of glycine it follows that at a pH of approximately 2, one hydrogen ion is split off. The resulting particle has a high dipole moment which can be explained solely by the presence of ionic charges in the amino acid molecule. A particle with the structure NH_2CH_2COOH could not have such a high dipole moment. Furthermore, we cannot expect that the substitution of the CH_2COOH group into the ammonia molecule, NH_3, which is a base of medium strength ($pK = 9.2$), would lead to the formation of an extremely weak base with $pK \approx 2$. The only explanation of these two facts is based on the assumption that the protolytic reaction

$$NH_3{}^+CH_2COOH + H_2O = NH_3{}^+CH_2COO^- + H_3O^+$$

takes place at pH 2. The ion $NH_3^+CH_2COO^-$ is called an *amphion* or *zwitterion* (German *zwitter* = hermaphroditic). This ionic form is stable in aqueous solutions and the tautomeric reaction

$$NH_3^+CH_2COO^- = NH_2CH_2COOH$$

almost does not take place. At pH as high as 9 a further protolytic reaction ensues:

$$NH_3^+CH_2COO^- + H_2O = NH_2CH_2COO^- + H_3O^+$$

Thus, the carboxylic group of glycine is more acidic than that of acetic acid because of the repulsive action of the positively charged NH_3^+ group on the hydrogen ion present in the carboxylic group. The $-NH_2$ group is more basic in glycine than in ammonia because the carboxylate group, $-COO^-$, electrostatically stabilizes the hydrogen ion in the ammonium group, $-NH_3^+$.

Consequently, the zwitterion $NH_3^+CH_2COO^-$ with a positive as well as a negative charge is the completely prevailing form of glycine in the pH range between 4 and 8. This range is termed *isoelectric* (Greek *isos* = equal). Analogous behaviour to that of glycine is exhibited by all amino acids that contain a single carboxylic and a single amino group as acidic or basic functional groups.

A slightly different situation arises with histidine

$$
\begin{array}{c}
N\!\!-\!\!C\!\!-\!\!CH_2\!\!-\!\!CH\!\!-\!\!COO^- \\
\|\quad\| \qquad\qquad | \\
CH\ \ CH \qquad\ NH_3^+ \\
\diagdown N \diagup \\
H
\end{array}
$$

which contains the weakly basic imidazolyl group (pK = 5.8). The pK of the ammonium group is 9.2. In the pH range where the amphion is the prevailing form of histidine, the presence of positively charged ions

$$
\begin{array}{c}
N\!\!-\!\!C\!\!-\!\!CH_2\!\!-\!\!CH\!\!-\!\!COO^- \\
\|\quad\| \qquad\qquad | \\
CH\ \ CH \qquad\ NH_3^+ \\
\diagdown \overset{+}{N} \diagup \\
H \qquad H
\end{array}
$$

or negatively charged ions

$$
\begin{array}{c}
N\!\!-\!\!C\!\!-\!\!CH_2\!\!-\!\!CH\!\!-\!\!COO^- \\
\|\quad\| \qquad\qquad | \\
CH\ \ CH \qquad\ NH_2 \\
\diagdown N \diagup \\
H
\end{array}
$$

cannot be neglected. Thus, the isoelectric pH range shrinks to an isoelectric point (pH$_{iso}$), where the concentration of positively charged ions is equal to that of negatively charged ions. From a simple calculation it follows that

$$pH_{iso} = \tfrac{1}{2}(pK_2 + pK_3)$$

where pK_2 and pK_3 are the dissociation constants of the ammonium group of the imidazolyl and of $NH_3{}^+$. In the isoelectric pH range or in the neighbourhood of the isoelectric point the solubility of the amino acids is lowest and they do not exhibit any mobility in an electric field.

Peptides are formed from amino acids by the peptidic bond

$$\begin{array}{cc} O & H \\ \| & | \\ -C & -N- \end{array}$$

If they are derived from amino acids that have one amino and one carboxylic group, the resulting chain includes ionizable groups only at both ends. In the peptidic chain of natural polymeric peptides, the *proteins*, there frequently occur amino acids originally possessing more than two ionizable groups, like histidine, glutamic and aspartic acids with one more carboxylic group, lysine with a further amino group, arginine with a guanidine group, etc. The polymer then contains a large number of ionized or ionizable groups and is termed a *polyelectrolyte*. Polyelectrolytes also include *nucleic acids* which, in a neutral solution, have a negative charge on every nucleotide unit owing to the dissociation of the phosphate group (see Figure 26). In addition, purine and pyrimidine bases contribute to the ionization equilibrium of nucleic acids in acidic solutions. Many polyelectrolytes are of a synthetic nature (for some examples see Table 7).

The characteristic phenomenon connected with polyelectrolytes in which they differ from low-molecular-weight electrolytes is the considerable electrical charge density in the space enclosed by the macromolecules (e.g. in the form of an α-helix or disordered coil; see Figure 28) and in its immediate surroundings. This space charge does not depend on the concentration of the polyelectrolyte. With respect to the electroneutrality condition in this space, there must be a concentration of 'small' ions corresponding to the number of opposite charges fixed in the structure of the polymer. Those ions are termed counterions or gegenions (German *gegen* = against). In the neighbourhood of the ionizable group those ions form a complete analogy of the ionic atmosphere (see page 35). When these ionic clouds come close together (where there are a sufficiently high number of ionized groups in the macromolecule), they repulse each other and the polyelectrolyte chain expands. For example, this effect contributes to the distortion (*denaturation*) of the double helix of deoxyribonucleic acid (DNA), which occurs when its stability is decreased by increasing the temperature. When a sufficient amount of low-molecular-weight electrolyte is added to the polyelectrolyte solution, it penetrates even to regions where only ionized groups and gegenions were originally present. Consequently, a certain concentration of ions with the same sign as the charges fixed in the polymeric chain is formed (they are termed *coions*). The increased electrolyte concentration in the surroundings of the macromolecule results in a decrease in the dimensions of the ionic atmospheres and in suppressing the repulsive influence of the gegenions. For example, in this way the stability of the double helix of deoxyribonucleic

54

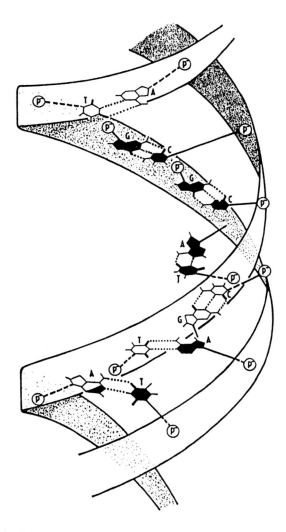

Figure 26. The double helix of DNA. Two helical bands represent the sugar–phosphate backbone of the structure. P⁻ denotes the phosphoric acid unit which is bound through the ester bonds to the neighbouring molecules of the sugar, deoxyribose. The remaining OH group is ionized. The inner part of each chain consists of a sequence of four bases: adenine (A), cytosine (C), guanine (G) and thymine (T). There are two possible links of the bases from one chain to the other: either adenine is bound to thymine by two hydrogen bonds or guanine is paired to cytosine by three hydrogen bonds (dotted lines). The bases are linked to the backbone by thick lines, either full (when the bond is seen) or dashed (when the bond is hidden by the backbone). In neutral media, which are actually present in organisms, the bases are not ionized. Thus, in a solution of low ionic strength the repulsing forces between the negative charges present on the phosphate units would distort the structure, with subsequent denaturation. However, a sufficient electrolyte concentration present in the organism suppresses the electrostatic field between the anion groups

Table 7. Examples of polyelectrolytes with acidic or basic function

Polyacrylic acid	$-CH_2-CH-$ $\quad\quad\;\; | $ $\quad\quad\; COOH$
Polystyrenesulphonic acid	$-CH_2-CH-$ (phenyl ring) SO_3H
Copolymer of styrene with maleic acid	$-CH_2-CH-CH-\!\!-\!\!-\!\!-CH-$ (phenyl) $\;\; COOH \quad COOH$
Polymetaphosphoric acid	$\quad\quad O$ $\quad\quad \uparrow$ $-O-P-$ $\quad\quad \|$ $\quad\quad OH$
Polyvinylpyridine	$-CH_2-CH-$ (pyridine ring, N)
Poly-4-vinyl-N-dodecylpyridinium (a 'polymeric soap')	$-CH_2-CH-$ (pyridinium ring, $\overset{+}{N}$) $(CH_2)_{11}-CH_3$
Copolymer of acrylic acid with 4-vinylpyridine (a polyampholyte)	$-CH_2-CH-CH_2-CH-$ $\quad\quad\;\; \|$ $\quad\quad\; COOH$ (pyridine ring, N)

acid is increased so that a considerably higher temperature is required for its denaturation than without the addition of the excess low-molecular-weight electrolyte.

When alkali metal hydroxide is continuously added to a solution of a polyelectrolyte that exhibits weakly acidic or ampholytic properties (i.e. the substance is titrated), the acid groups ionize. For example, in polyacrylic acid the hydrogen ion is initially bonded with equal strength to each carboxylic

56

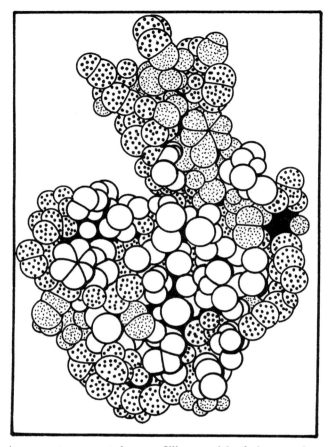

Figure 27. A computer-generated space-filling model of the protein egg lysozyme, showing the largest section through the molecule. The void circle is the hydrophobic unit, the finely dotted circle is the slightly polar unit and the coarsely dotted circle is the strongly polar unit. (According to G. D. Rose)

group. When the macromolecule ionizes, the resulting negative charge prevents further splitting off of hydrogen ions so that the acidity of the remaining carboxylic group decreases. In this situation each carboxylic group will possess a definite effective pK value which is higher than that of monomeric acrylic acid. This change of pK helps to elucidate the dependence of the pH of a solution, e.g. of a protein, on the amount of hydroxide added (the titration curve).

Let us consider the behaviour of naturally occurring polyelectrolytes, proteins and nucleic acids, in more detail.

The sequence of amino acids in the individual peptide chains of a protein is called the *primary structure* of the protein. Hydrogen bonds between amino acid units influence the configurations of the individual chains. These *secondary structures* (cf. page 29) have the form of an α-helix or a β-structure which resembles a pleated leaf consisting of stretched polypeptide chains. The *tertiary*

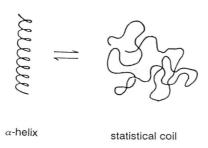

α-helix   statistical coil

Figure 28. Transition of an α-helix to a statistical coil

structure, i.e. the mutual arrangement of the chains, is governed for a major part by hydrophobic interactions between low-polarity groups in the protein molecule. A typical example of this behaviour are globular proteins, the ordered coil of which has a core held together by hydrophobic interactions (see Figure 27).

In an aqueous solution the surface of the protein is hydrated and water molecules fill the cavities inside the protein structure.

Strongly alkaline or strongly acidic media also have a strong impact on the behaviour of proteins. Under these conditions the acidic or basic groups in the macromolecule ionize to such an extent that an α-helix–disordered coil transition takes place (see Figure 28). This process is called protein denaturation. A protein can also be denatured by an elevated temperature (the helix breaks down due to excessive thermal vibration of the chain) or by the addition of urea (the hydrophobic interaction holding an ordered structure together ceases to exist). A denatured protein loses its functions (e.g. an enzyme becomes inactive) and its solubility in water decreases so that coagulation takes place.

In contrast to proteins, which have a more or less ordered but considerably irregular structure, the double helix of DNA is an example of a regular structure (see Figure 26). The uncompensated charges of the phosphate groups are screened off by ionic atmospheres built of gegenions. The structure is reinforced by hydrogen bonds between the nitrogen bases and by van der Waals interactions between their planar rings. In the presence of magnesium ions which form ion-pairs with the phosphate groups the structure is even more stabilized. The whole structure is hydrated by about 20 water molecules per nucleotide. As already mentioned, the DNA structure can resist considerable pH changes and its denaturation can be brought about at elevated temperatures. In a similar way as in the case of proteins the hydrophobic interactions may be distorted by urea with subsequent denaturation.

References

1. H. Morawetz, *Macromolecules in Solution*, John Wiley & Sons, New York, 1965.
2. F. Oosawa, *Polyelectrolytes*, Marcel Dekker, New York, 1971.
3. *Prediction of Protein Structure and the Principles of Protein Conformation* (G. D. Fasman, ed.), Plenum Press, New York, 1989.
4. *Biological Macromolecules and Assemblies*. Vol. 2. *Nucleic Acids and Interactive Proteins* (F. Jurnak and A. McPherson, eds.), John Wiley & Sons, New York, 1985.

PARTICLE MOTION IN QUIET SOLUTIONS

Two glass vessels 1 and 2 containing potassium chloride solutions with two different concentrations, c_1 and c_2, are connected by a thin short glass capillary (Figure 29). The solutions in both vessels are stirred so that the concentration of potassium chloride is constant in each vessel. However, the stirring has no effect in the capillary. The concentration changes are measured in both vessels by a suitable sensor (e.g. a potassium ion-selective electrode, see page 144, or, still better, by measurement of the conductivity in vessel 2). There is some delay before a uniform drop of concentration of potassium chloride in the capillary is established between values c_1 and c_2 and then the measurement can begin. The results show that the rate of penetration of the electrolyte, v, is directly proportional to the concentration difference between the two vessels and inversely proportional to the length of the capillary, l,

$$v = \frac{k(c_1 - c_2)}{l} \tag{1}$$

In a different electrolyte the same dependence is obtained with a different value of the proportionality constant k. The only influence acting on the motion of the ions is the concentration difference (consequently, the osmotic pressure difference); the stirring of the solutions plays no role. The process of transport of a substance controlled by the concentration difference of that substance is called *diffusion*. It is governed by the *Fick law*, which can be deduced by generalization of equation (1). The rate of any transport process (not only diffusion) is characterized by its flux, which is the amount of substance (in moles) or another quantity, e.g. heat, that passes through a unit area (1 m² or 1 cm²) during 1 second. When the concentration changes only along a single coordinate, x (the linear diffusion), the Fick law states that the flux is proportional to the *concentration gradient*:

$$J = -D \text{ gradient } c$$

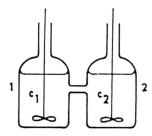

Figure 29. A device for the steady-state diffusion. Cells 1 and 2 are filled with an electrolyte solution of concentrations c_1 and c_2. A constant concentration is maintained in each cell by stirring. Diffusion is restricted to the connecting capillary only

where gradient c is the ratio of the increment of concentration Δc and of the increment of distance Δx (gradient $c = \Delta c/\Delta x$; in a more exact formulation, $\Delta c/\Delta x$ is to be replaced with the differential quotient dc/dx). In the present case,

$$\Delta c/\Delta x = \frac{c_2 - c_1}{l}$$

and it must be kept in mind that the concentration gradient corresponds to the change of concentration with *increasing* coordinate. The opposite term is *concentration drop*, which biologists often confuse with the concentration gradient.

The proportionality constant D (with units of square centimetres per second when the unit of flux is moles per square centimetre per second and the unit of concentration is moles per cubic centimetre) is called the diffusion coefficient, or diffusivity, and specifies the diffusion rate of the given substance. The minus sign shows that the diffusion flux is oriented from higher to lower concentrations.

In a physical or chemical process a characteristic quantity (e.g. concentration or temperature) usually changes with time (e.g. the concentration of a reactant during a chemical reaction or the temperature of a body during heating) so that the change of this quantity with respect to time, dc/dt, is different from zero. In certain cases, however, the process approaches a steady state, so that the rate of change of the characteristic quantity (usually a function of the coordinates of the system) approaches zero. This situation prevails in our experiment: a steady concentration distribution is established in the capillary, where the concentration decreases linearly with distance. In other words, the same quantity of potassium chloride that came from vessel 1 into the capillary leaves it by diffusion into vessel 2.

However, many diffusion processes do not take place at steady state but have a transient nature, i.e. the concentrations are time-dependent. In the next experiment the vessel (see Figure 30) is filled with a solution of a coloured salt, e.g. potassium permanganate. A long capillary containing pure water is attached to the vessel. The solution in the vessel is stirred so that the concentration of potassium permanganate is constant everywhere in the vessel, while the stirring

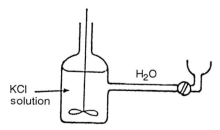

Figure 30. A device for the study of linear diffusion in an unsteady state. The solution in the cell is stirred so that its concentration is constant in the whole cell. In the capillary the solute diffuses into pure water

60

has no effect in the capillary. Since the volume of the capillary is negligible compared with the volume of the vessel, the quantity of potassium permanganate penetrating by diffusion has virtually no influence on the overall composition of the solution in the vessel. The concentration of potassium permanganate at any distance from the orifice of the capillary is measured with a special microspectrophotometer. The capillary is rather long so that the penetration of potassium permanganate is observed only in the region close to the vessel. Thus, the principal features of the diffusion process in the capillary are the same as the physical model of linear diffusion in an infinitely long tube.

When the concentration of potassium permanganate is measured at different distances from the orifice as a function of time, the dependence shown in Figure 31 is obtained. Obviously, the slope of the concentration at the origin of the coordinate system decreases with time, i.e. the diffusion rate diminishes. The concentration gradient for $x = 0$ is described by the equation (which is deduced theoretically but corresponds exactly to our experiment)

$$\text{gradient } c \text{ (for } x = 0) = \frac{c^0}{(\pi D t)^{1/2}}$$

where c^0 is the concentration of potassium permanganate in the cell and t is the time elapsed from the start of the experiment when the capillary came into contact with the solution in the vessel. In Figure 31 we observe that the region into which potassium permanganate has penetrated expands with time. This region is called the diffusion layer. Since this layer has, of course, no definite border we must define a quantitative measure for its effective thickness. A

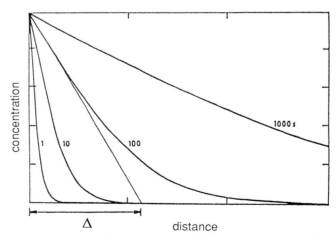

Figure 31. Concentration distribution in the capillary as a function of distance from its orifice into the cell (cf. Figure 30) for different values of time elapsed from the start of the diffusion. The effective diffusion layer thickness Δ is indicated on the diagram for a time of 100 s

tangent is constructed to the concentration curve at point $x = 0$. According to the above equation the distance of its intersection with the line $c = 0$ (x axis) from the coordinate origin is

$$\Delta = (\pi Dt)^{1/2}$$

The length Δ is termed the effective diffusion-layer thickness. The result of a calculation indicates that, at distance $x = \Delta$ from the orifice of the capillary, the concentration of the diffusing substance decreases to approximately 21% of the original value. This decrease is independent of the diffusivity as well as of the concentration. The effective diffusion-layer thickness gives an illustrative picture of the rate of propagation of a substance by diffusion.

So far a macroscopic picture of diffusion has been considered. Continuous distribution of the content of the substances in the solvent medium has been assumed and the molecular point of view, i.e. the behaviour of an individual particle in the solution, has been ignored. The former approach is termed phenomenological.

When the motion of individual particles is considered it cannot be assumed that they move straight in the direction of the drop in concentration. In fact, they are located in the cavities between the solvent molecules, where they vibrate. After several hundred vibrations they acquire sufficient energy in a random event to jump into a neighbouring stable position where the same process occurs again. In this way the ions move irregularly, as shown in Figure 32. As a result of this random motion the particle travels to an average distance of about several hundredths of a millimetre in a second. Einstein and Smoluchowski worked out the statistical theory of diffusion and, in connection with this theory, deduced the expression for the average distance $\bar{\Delta}$ that a particle covers during the observation time τ:

$$\bar{\Delta} = \sqrt{2D\tau}$$

where D is the diffusion coefficient of the particle (see Figure 32).

When the solution contains a constant concentration of the solute this translocation of particles is of no importance for any total change. (In this

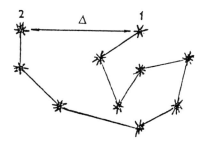

Figure 32. Irregular particle motion in solution. The particles oscillate about a mean position for some time and then jump to a neighbouring position. Δ denotes the displacement of the diffusion particle during the time τ

paragraph the diffusing species are termed particles because the discussion is the same for ions as for uncharged molecules.) When the particle leaves position 1 another particle subsequently moves to this position. However, new circumstances prevail when a concentration gradient is formed in the solution. The probability of a particle moving in the direction of increasing concentration is the same as in the direction of decreasing concentration. However, the probability for particles to be transported from the region of higher concentrations is greater than for those coming from the region of lower concentrations. This results in the prevalent transfer of the solute in the direction of the concentration drop, as described by the Fick law.

References

1. Page xiii, Ref. 2, Chapter 2.
2. J. Crank, *The Mathematics of Diffusion*, Clarendon Press, Oxford, 1964.

PARTICLE MOTION IN STIRRED SOLUTIONS

Why have thin capillaries been used for the preceding experimentation instead of thick tubes? The answer is simple. In such a case the transport of the substance would take place in a non-uniform way, which is obvious a few seconds after the start of the experiment. In a thick tube even the small local differences of temperature and concentration cause a non-uniform distribution of the density of the solution, i.e. density gradients. Through the action of the force of gravity the denser regions move downwards and push the thinner regions upwards which leads to streaming of the solution, called *natural convection*. This phenomenon which accelerates the transport plays an important role in nature.

The transport rate can also be increased by the experimenter, e.g. by stirring. In this case the mechanical impulse is transferred from the stirrer to the liquid medium.

The transport simultaneously mediated by convection and diffusion is called *convective diffusion*.

Let us consider a plate made of a soluble substance which is placed in a solvent which flows with a certain velocity V (in cubic centimetres per second). If we assume that the dissolving rate of the substance is high (which is not always the case) a saturated solution of the substance is formed at the surface of the plate. It is transported to the bulk of the solution by convective diffusion. However, the flow of the solvent is slowed down by the plate. Directly at the surface its velocity is zero while it increases with increasing distance from the plate until it reaches the value V (see Figure 33). The decelerating effect of the solid plate is due to *inner friction* (*viscosity*) of the liquid. The completely immobile layers of the liquid in the immediate vicinity of the plate slow down the more distant layers. This influence decreases in the direction towards the

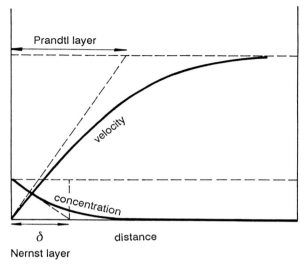

Figure 33. Dependence of the velocity of the liquid and of the concentration of the dissolved substance on the distance from the plane surface of a crystal during its dissolution and solute transport by convective diffusion. The Prandtl (hydrodynamic) and the Nernst layers characterize the distribution of velocity and concentration

bulk of the solution. The region of decreasing velocity of the solution is called the *Prandtl* or *hydrodynamic layer*. Convective diffusion occurs in a thinner layer, the so-called *Nernst layer* (Figure 33). It preserves its transient feature only at the beginning of the transport process and soon turns into a steady state characterized by the formation of the Nernst layer of time-independent thickness. The steady state ensues because the penetration of the dissolved substance into the bulk of the liquid is compensated by the influx of the solvent into the vicinity of the plate which is more effective at larger distances from the surface. The flux of the substance characterizing the dissolving rate at a unit area of the plate is given by

$$ J = \frac{Dc_s}{\delta_N} $$

where c_s is the concentration of the saturated solution and δ_N the Nernst layer thickness. The flux soon becomes larger than the supposed flux due to 'quiet' diffusion.

A similar situation is also encountered in electrochemical systems (page 106). In cell biology the immobile (stagnant) layers at the surface of cells are often mentioned. In the case of tissues this concept is not far from reality because in the thin layer of intercellular liquid no convection can occur because of internal friction of the liquid (in other words, the Prandtl layer due to natural convection would be much thicker than the thickness of the intercellular space). A similar situation ensues in a thin capillary where 'quiet' diffusion takes place (see page

64

59). In contrast, during the motion of organisms in water or at the surface of tissues the Prandtl layer must appear with continuously decreasing velocity of the liquid.

References

1. E. L. Cussler, *Diffusion: Mass Transfer in Fluid Systems*, Cambridge University Press, Cambridge, 1984.
2. V. G. Levich, *Physico-Chemical Hydrodynamics*, Prentice-Hall, Englewood Cliffs, N.J., 1962.
3. J. Newman, *Electrochemical Systems*, Prentice-Hall, Englewood Cliffs, N.J., 1973.
4. Page xiii, Ref. 6, Volume 6.

ION MOVEMENT IN ELECTRIC FIELDS

The first experiment described on page 58 will now be modified somewhat. As shown in Figure 34, into each of vessels 1 and 2 is placed a silver–silver chloride electrode which is a silver plate covered with insoluble silver chloride. More will be said about the properties of this electrode on page 87. Both electrodes are connected to a d.c. voltage source. This source can be, for example, a potentiometer connected to a storage battery. The voltage that is fed to the electrodes from the potentiometer depends on the position of the sliding contact or the potentiometric wire (the resistance wire, usually made from an iron–nickel alloy has a constant cross-section). The resulting voltage is given by the relationship

$$\Delta V = U_{st} \frac{\overline{AS}}{\overline{AB}}$$

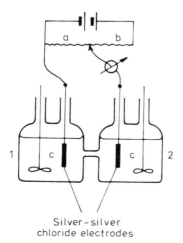

Silver–silver chloride electrodes

Figure 34. A device for the study of ion migration. In both cells the electrolyte concentration is the same and migration occurs in the connecting capillary only

where $\overline{AS}$ is the length of the segment of the potentiometric wire between its origin and the position of the sliding contact, $\overline{AB}$ is the total length of the potentiometric wire and U_{st} is the voltage of the storage battery. The electrical circuit also includes a galvanometer G which measures the electrical current flowing between the voltage source, the vessels and the connecting capillary. Silver–silver chloride electrodes are used to obtain an electrical potential difference between the electrodes in the solution that is just equal to the voltage supplied by the potentiometer and to prevent electrode polarization (loss of voltage at electrodes under current flow, seen in detail on page 104). In contrast to the experiment described earlier, both vessels and the capillary are filled with a potassium chloride solution of the same concentration.

The current flowing between the two compartments is directly proportional to the voltage imposed on the electrodes and to the potassium chloride concentration (at least at lower concentrations), inversely proportional to the length of the capillary and directly proportional to its cross-section. The current increases with increasing temperature of the electrolyte. When a positive charge Q has passed from vessel 1 to vessel 2, the increase in the content of potassium chloride in vessel 2 is $0.49Q/F$, while the content of potassium chloride in vessel 1 has decreased by the same amount (the amount of potassium chloride in the capillary is negligible). The quantity $F = 96\,480$ C/mol is again the Faraday constant (page 35), corresponding to a charge of 1 mol of univalent cations.

On the basis of these observations it can be concluded that the flow of electric current through an electrolyte follows Ohm's law and, therefore, the electrolyte conductivity is independent of the voltage fed to the electrodes. The system of two vessels joined by a thin capillary with a voltage source connected to the electrodes can be represented by the substitution diagram shown in Figure 35.

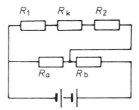

Figure 35. A substitution diagram for the device in Figure 34. The resistance of the solution in the capillary is much larger than that of cells 1 and 2

The resistances of the compartments, R_1 and R_2, are much smaller than the resistance of the capillary, R_c. The entire voltage of the source therefore acts between the ends of the capillary. Thus, the length and cross-section of the capillary determine the resistance of the whole system, as has already been observed.

The change in the amount of potassium chloride contained in each vessel can be explained in the following way. When a positive charge is injected from the silver–silver chloride electrode into the solution in vessel 1, the reaction

$$Ag + Cl^- = AgCl + e$$

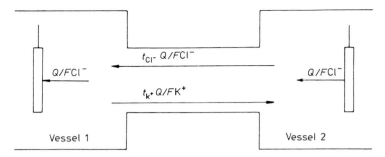

Figure 36. Charge transport by cations and anions in a tube

takes place at the electrode. Thus, instead of a positive charge passing into the solution, a negative charge passes from the solution into the electrode. Nevertheless, the result is the same: the amount of chloride in vessel 1 decreases by Q/F (see Figure 36). The fact that the decrease in the potassium chloride content in vessel 1 is only $0.49Q/F$ can be explained by the transport of $0.51Q/F$ mol of Cl^- from vessel 2 to vessel 1. At the same time the above reaction occurs at the electrode of vessel 2 in the opposite direction. However, the increase in the amount of potassium chloride in vessel 2 is only $0.49Q/F$ because just this amount of potassium ions has passed from vessel 1 to vessel 2 and $0.51Q/F$ of chloride ions has been transported in the opposite direction. If our observation is restricted to the processes in the capillary, the total charge Q is transported partly in the direction from vessel 1 to vessel 2 by cations as the fraction $0.49Q$ and partly in the opposite direction by anions as $0.51Q$. If the charge were also carried, for example, by electrons (fraction xQ) the amount of charge transported by the potassium ions would decrease to $0.49(1-x)Q$ and that transported by chloride ions to $0.51(1-x)Q$ which, of course, is not in accord with observation.

Thus, since the electrical charge is transported only by potassium and chloride ions, the electrolyte conductivity is directly proportional to the concentration of potassium chloride (if this is not excessively high). In contrast to a metallic conductor, the conductivity of an electrolytic conductor increases with temperature, because at higher temperatures the ions move more rapidly. The mobility of various types of ions is different. As we have already seen, in a potassium chloride solution the potassium ions transport 49% of the overall charge transferred. The ratio between the charge transported by an individual type of ion to the total charge transferred is called the transport number t_i. Thus, $t_{K^+} = 0.49$ and $t_{Cl^-} = 0.51$. Potassium chloride is an exceptional case where the transport numbers of the cation and of the anion are almost equal. On the contrary, for lithium chloride, for example, $t_{Li^+} = 0.33$ and $t_{Cl^-} = 0.67$.

The transport numbers reflect the relative speed of the ions. We will now consider the absolute values of ion mobilities. The electric current flowing through the capillary is related to the current density j (with units of current per unit area, e.g. A cm^{-2}) by the equation

$$I = Aj$$

where A is the cross-section of the capillary. The current density is composed of the flux of cation J_B and of anion J_A:

$$j = F(J_B - J_A)$$

The minus sign indicates that the anions move in an electrical field in the opposite direction to the cations. This equation is a formulation of the *Faraday law* which governs the electrolytic charge transport or migration (i.e. the transport of charge linked to the transport of matter). In a more general formulation, when the charge number of the cation is z_B and that of the anion z_A,

$$j = F(z_B J_B + z_A J_A)$$

The flux of an ion is directly proportional to the ion concentration at a given point, to the charge number and to the 'driving force' of the transport which is, in the present case, the electrical field strength. The resulting product must be multiplied by the Faraday constant to convert quantities related to the transport of electricity to the transport of matter. Here the electrical field strength X is simply equal to the ratio of the voltage or, more properly, the electrical potential difference, acting between the ends of the capillary and its length $\Delta V/l$, multiplied by -1. The fluxes of the cation J_B and of the anion J_A are then given by the equations

$$J_B = u_B F c_B z_B X = -u_B F c_B z_B \frac{\Delta V}{l}$$
$$J_A = u_A F c_A z_A X = -u_A F c_A z_A \frac{\Delta V}{l}$$

(1)

The proportionality constants u_B and u_A are the mobilities of ions B and A. The mobilities reflect the ease with which a particle overcomes the resistance of the medium to its movement.

In general, the electric potential φ is the electric work necessary for transfer of a unit charge, for example 1 coulomb, from infinity to a given site. The ratio of voltage and length of the tube will be substituted by a more general term, which is the gradient of the electric potential $\Delta\varphi/\Delta x$. Equation (1) will be transformed, for example for cation B, to the form

$$J_B = -u_B F c_B z_B \text{ gradient } \varphi$$

(2)

The mobility u characterizes the velocity of penetration of a particle through the medium under the influence of any physical field while the quantity specific for ions is the electrolytic mobility, given by the product of the charge of 1 mol of ion i (regardless of its sign), $|z_i| F$, and of the mobility; for example, in the case of a cation,

$$U_+ = z_+ F u_+$$

Table 8. Ion conductivities

Ion	Ion conductivity ($S \ cm^2 \ mol^{-1}$)	Ion	Ion conductivity ($S \ cm^2 \ mol^{-1}$)
H_3O^+	349.8	OH^-	198.3
Li^+	38.7	F^-	55.4
Na^+	50.1	Cl^-	76.4
K^+	73.5	Br^-	78.1
Rb^+	77.8	I^-	76.8
Cs^+	77.3	SO_4^{2-}	160.0
Ca^{2+}	119.0	$Fe(CN)_6^{3-}$	302.7

In electrochemical tables ion conductivities defined as $\lambda_i = z_i^2 F^2 u_+$ are often listed (for example, see Table 8). Ion conductivities of components of a given electrolyte decide how the cation and the anion contribute to current flow at their unit concentration. As already mentioned, the share of each ion is quantitatively expressed by its transport (transference) number:

$$t_+ = \frac{\lambda_+}{\lambda_+ + \lambda_-}, \qquad t_- = \frac{\lambda_-}{\lambda_+ + \lambda_-}$$

In an analogous way as in the case of equilibria in electrolyte solutions the transport properties are affected by electrostatic and other forces originating in the surroundings of the moving ion. The most striking effect is produced by the ionic atmosphere which, in an electric field, moves in a direction opposite to that of the ion. The necessity to rebuild the ionic atmosphere around the escaping ion is equally important.

The velocity of a transport process (diffusion, ion migration and convection, but also, for example, heat conduction, i.e. transport of thermal energy) depends on the species transported and on the kind of force that drives the particular kind of transport. In the theory of transport processes (which is a part of irreversible thermodynamics) there appear various *driving or phenomenological forces*. They characterize the deviation of the system from equilibrium. The larger this deviation the stronger the tendency of the system to return to the equilibrium situation. This deviation from equilibrium can be mostly expressed quantitatively as the local change of the Gibbs (free) energy, the amount of energy that can be converted ('transduced') to other sorts of useful energy, such as mechanical or electrical work, or energy stored in an 'energy-rich' compound (cf. pages 166 and 184).

In chemical systems and, to a degree, in biological systems, the Gibbs energy can be distributed among the components of the system. The *chemical potential* of an uncharged component of a system is the amount of Gibbs energy inherent in 1 mol of that component. In the case of a dilute solution the chemical potential of a component i is

$$\mu_i = \mu_i^0 + RT \ln c_i$$

Here μ_i^0 denotes the standard chemical potential* and c_i the concentration of that component; μ_i^0 is independent of concentration c_i. If, for example, the chemical potential has different values between individual points in the solution the system is not in equilibrium and it is the gradient of the chemical potential by which the tendency of the system to move towards equilibrium is expressed. Thus, this gradient is the driving force of diffusion.

In an electrolyte solution, the electrical energy of i,

$$\mu_{e,i} = z_i F \varphi$$

must be added to the chemical potential in order to obtain a complete Gibbs energy term called the electrochemical potential:

$$\tilde{\mu}_i = \mu_i + \mu_{e,i} = \mu_i^0 + RT \ln c_i + z_i F \varphi \tag{3}$$

This important concept will often appear in the present book. In fact, we have already met it: the physicists use a different name for it, the Fermi level of electron energy (see page 12).

The transport process characterized by the gradient of the electrochemical potential is simultaneously occurring diffusion and ion migration (sometimes called electrodiffusion for both processes together). The rate of this transport process can be expressed simply as the flux of the substance proportional to the number of particles (or concentration c_i) and to the driving force with the mobility of the substance u_i as the proportionality constant:

$$J_i = -u_i c_i \text{ gradient } \tilde{\mu}_i$$
$$= -u_i RT \text{ gradient } c_i - u_i z_i F \text{ gradient } \varphi$$
$$\left(= -u_i c_i \frac{d\tilde{\mu}_i}{dx} \right.$$
$$\left. = -u_i RT \frac{dc_i}{dx} - u_i z_i c_i F \frac{d\varphi}{dx} \right)$$

This relationship is called the Nernst–Planck equation. Obviously it is a combination of the Fick law (page 58) and the equation of ion migration (equation (2), page 67). The diffusion coefficient D_i is obviously related to the mobility u_i by the equation

$$D_i = RT u_i \tag{4}$$

---

* The values of standard chemical potentials may be found in standard textbooks of thermodynamics and in tables of physicochemical constants under the name standard molar Gibbs energies.

On introducing the electrolytic mobility $U_i = |z_i| F u_i$ (cf. page 67) we obtain

$$\frac{D_i}{U_i} = \frac{RT}{z_i F}$$

which is the Nernst–Einstein equation, valid, however, only for dilute solutions.

The Nernst–Planck equation relates the diffusion and the ion migration by a single constant, u_i, characterizing the individual properties of a given type of ion. At this stage we shall be able to compare the rates of the two processes. Let us return to the experiment on steady-state diffusion (Figure 29) and ion migration (Figure 34). In the diffusion experiment 10^{-3} M KCl is in compartment 1 and 2×10^{-3} M KCl in compartment 2. In the migration experiment the same concentration, 10^{-3} M KCl, is in both compartments. The length of the capillary is 0.5 cm and the temperature of the solutions is 25 °C (298 K). For the sake of simplicity it is assumed that both the diffusion coefficient of K^+ and of Cl^- have the same value, i.e. 2×10^{-5} cm^2 s^{-1}. In order to obtain the result in coherent units we must recalculate the concentrations to units of moles per cubic centimetre (1 M KCl = 10^{-3} mol cm^{-3} KCl). From the equation (cf. page 58)

$$J = -D \frac{c_1 - c_2}{l}$$

we have for the diffusion flux $J_{KCl} = -4 \times 10^{-11}$ mol cm^{-2} s^{-1}. Now let us determine the magnitude of the voltage that must be imposed on the silver–silver chloride electrodes to obtain the same migration flux of potassium ions (the flux of chloride ions has the same absolute value but the opposite sign). Equations (1) and (4) yield the relationship

$$\Delta V = \frac{l J_{K^+}}{u_{K^+} F c_{K^+}} = \frac{RT J_{K^+}}{D_{K^+} c_{K^+} F} = 0.026 \text{ V}$$

Thus, in the absence of extreme concentration gradients in the solution, the transport by ion migration at electrical field strengths of an order of volts per centimetre is much faster than diffusion.

In the next experiment the experimental arrangement shown in Figure 29 will again be used, but compartments 1 and 2 will be filled with an electrolyte solution whose cation and anion have very different transport numbers. Thus, vessel 1 contains 10^{-4} M HCl and vessel 2 contains 10^{-3} M HCl. Once diffusion has attained the steady state, the electrical potential difference between the vessels is measured (e.g. by two silver–silver chloride electrodes connected to each of the solutions by a capillary tube filled with a saturated KCl solution; see page 87). The voltage between the electrode placed in vessel 2 and the electrode in vessel 1 measured by an electronic voltmeter is approximately -38 mV (the electrode placed in vessel 2 being negative with respect to the electrode placed in vessel 1).

Thus, electrolyte diffusion alone can produce electrical potential differences in the solution. Obviously, ions of higher mobility (here H_3O^+) tend to outrun the ions with lower mobility (Cl$^-$). This cannot result in separate accumulations of cations or anions because electroneutrality must be preserved. However, the fast and the slow ions interact electrostatically to produce an electrical potential difference in the solution. This *diffusion potential* is the quantity we have measured. Diffusion potential is a general term for electrical potential changes accompanying electrolyte diffusion. In our experiment the vessels where the diffusion is negligible are connected with a junction where it actually takes place. The potential difference between the solutions is then called the liquid junction potential $\Delta\varphi_L$.

The solution of the Nernst–Planck equation for the case of absence of electric current gives

$$\Delta\varphi_L = \frac{RT}{F}\left(\frac{u_{H_3O^+} - u_{Cl^-}}{u_{H_3O^+} + u_{Cl^-}}\right)\ln\frac{c_2}{c_1}$$

$$= \frac{RT}{F}(t_{H_3O^+} - t_{Cl^-})\ln\frac{c_2}{c_1}$$

which, for 25 °C and using decadic logarithms, turns into

$$\Delta\varphi_L = 0.0591\,(t_{H_3O^+} - t_{Cl^-})\log\frac{c_2}{c_1}$$

The term $2.3\,(RT/F) = 0.0591$ V is called the Nernst coefficient.

The transport number of the oxonium ion in hydrochloric acid is $t_{H_3O^+} = 0.82$. Thus $\Delta\varphi_L = 38$ mV.

If the two vessels are connected with a tube with a saturated solution of potassium chloride in agar gel the value of $\Delta\varphi_L$ is about 1 mV. In this 'liquid bridge' the charge carriers are predominantly potassium and chloride ions because of their high concentration. As they have approximately equal mobilities no diffusion potential is formed. A practical arrangement for the removal of the diffusion potential between two electrolyte solutions is shown in Figure 37.

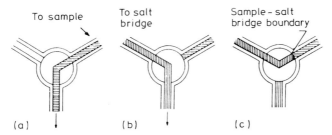

Figure 37. Liquid junction with free diffusion: (a) the test solution is drawn (in the direction of the arrow) into the stopcock; (b) the saturated KCl solution from the liquid bridge is drawn into the stopcock; (c) by turning the stopcock the test solution/liquid bridge liquid junction is formed. (According to G. Mattock and D. M. Band)

Removal of the diffusion potential by a liquid bridge is important for potentiometry, which belongs among very useful electroanalytical methods (see page 126).

In the next experiment the solution in vessels 1 and 2 will contain 0.1 M KCl as well as 10^{-4} M and 10^{-3} M HCl. The voltmeter connected to the silver–silver chloride electrodes indicates zero. Excess potassium chloride acting as an *indifferent electrolyte* (cf. page 37) prevents the formation of a diffusion potential. At any position in the electrolyte the ions of the indifferent electrolyte are present in a much larger concentration than the diffusing ions. They suppress the electrostatic interaction between the diffusing ions so that no appreciable electrical field is formed.

Let us now take a microscopic view at ion motion under electric forces. When an external source of electricity produces an electrical field in the solution, ion migration begins. Movement in the direction of the field or against it results from the slight preference among the cations for movement in the direction of the field (among the anions in the opposite direction). An electrical field of 1 V cm^{-1} will be considered which can easily be realized in the experiment shown in Figure 34. This field results in a component of the velocity of a cation in the direction of the field of about $5 \text{ } \mu\text{m s}^{-1}$. The velocities of the ions when they move from one stable position to another are approximately 10^7 times larger, so that when an ion makes about 10^{11} jumps per second the external electrical field produces a shift corresponding to only 10^4 jumps per second. This is a minute influence compared with the thermal motion of the ions (this situation is analogous to the conduction of electricity in a metal; see page 12). Nevertheless, the influence of the field on the oriented motion of the ions is quite marked.

The ion conductivity (see Table 8) and, therefore, also the diffusion coefficient of the hydronium ion are much larger than those of other ions, although the hydronium ion is not very much smaller. This striking phenomenon results from the fact that, in the translocation of the hydronium ion, the whole structure is not transferred—only an arbitrary proton from the hydronium ion:

$$\begin{matrix} H \\ \quad \\ H \end{matrix} \!\!\!\! \diagdown\!\!\diagup \!\! O\!-\!H^{+} \rightarrow O \!\!\! \diagup\!\!\diagdown \!\!\!\! \begin{matrix} H \\ \quad \\ H \end{matrix}$$

Proton transfer from one water molecule to another requires a certain activation energy. With light particles such as the electron, similar transfer processes occur by the tunnel mechanism: the electron wave function extends across the barrier which inhibits the electron transfer and, therefore, a certain probability exists that the transfer occurs without supplying activation energy to the electron (see also page 103). As shown by Bernal and Fowler, under certain conditions the proton can also be transferred by a tunnel mechanism (of course, to shorter distances than the electron). If, however, proton transfer in water occurred by a pure tunnel mechanism, the ion mobility and the diffusion coefficient would be much larger than the values found experimentally. Obviously the transfer rate is not in fact determined by the actual velocity of proton

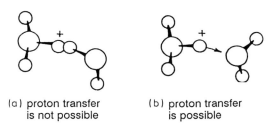

(a) proton transfer
is not possible

(b) proton transfer
is possible

Figure 38. Proton transfer from one water molecule to another requires suitable orientation of both molecules

motion, which is very large, but by the rotation of the water molecule to a position in which it can accept the proton (see Figure 38).

References

1. Page xiii, Ref. 2, Chapter 2.
2. C. A. Vincent, 'The motion of ions in solution under the influence of an electric current', *J. Chem. Educ.*, **53**, 490 (1976).

Chapter 2

Electrodes

OXIDATION AND REDUCTION AT ELECTRODES

Electrolysis has already been mentioned on page 6; here an experiment with electrolysis in a modified arrangement will be described. The *electrolytic cell* (Figure 39) will consist of a glass vessel containing a large-surface platinum electrode (e.g. a plate with an area of 1 cm^2) and a small platinum electrode (e.g. a platinum wire with a diameter of 0.1 mm sealed in a glass tube so that a length of 0.5 mm projects). The surface area of the small electrode (micro-electrode) is 1.6×10^{-3} cm^2. The electrolyte solution in the cell is 10^{-3} M Fe(ClO$_4$)$_2$, 10^{-3} M Fe(ClO$_4$)$_3$ and 0.1 M HClO$_4$. Both electrodes are connected to a suitable voltage source (cf. Figure 34). In the present case an electronically controlled low-output impedance voltage source is used instead of a simple potentiometer. The current-measuring device, which was a galvan-ometer in Figure 34, should record even rather rapidly changing current. Thus, a fast pen recorder or an oscilloscope must be used. These instruments record the electric current (vertical axis) as a function of time (horizontal axis).

When zero voltage ($\Delta V = 0$) is imposed on the electrodes, no current deflection is indicated by the recorder. Obviously, the system is in equilibrium. Assume that a small voltage (for example 5 mV) is applied to the electrodes, polarizing them so that the small electrode is positive. The recorder first indicates a very high and rapidly decaying positive current. Later the current drops rather slowly (Figure 40a). When the voltage source is switched off and the electrical potential difference between the electrodes is measured by a recording electronic voltmeter (no current flows through the instrument because of its high input resistance), the initial value of 4 mV falls slowly to zero (Figure 40b).

The time-course of the current during switching-on of the external voltage source (i.e. during electrolysis) as well as the time-course of the electrical potential difference between the electrodes after switching off the voltage source are caused by several processes that take place at the electrodes, in their surroundings and in the bulk electrolyte. Because of the much smaller size of the microelectrode compared with the large electrode, the current density at the

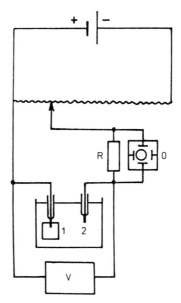

Figure 39. A simple electrolysis arrangement. A large-surface electrode (1) and a microelectrode (2) are polarized by a voltage source. The current is measured with a recording ammeter or oscilloscope as a function of time and the electric potential difference between the electrodes is measured with a recording voltmeter

microelectrode is many times (three orders of magnitude) smaller than at the other electrode. Therefore the microelectrode, where the effect of imposed voltage or polarization is apparent, is often called the *polarized* (or polarizable) electrode, while the large electrode where this effect is negligible is non-polarized.

As will be discussed in more detail below, the electrode has two basic properties. It is an acceptor or donor of electrons (or holes, eventually) which

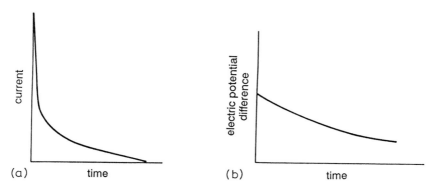

Figure 40. Dependence of (a) current and (b) potential difference between the electrodes on time in the experiment described by Figure 39

are transferred from the solution or into the solution. On the other hand, when it is in contact with an electrolyte solution, it bears a definite charge uniformly spread over its surface contacting the solution. The charge of the electrode attracts ions of opposite sign from the solution and, in this way, a molecular condenser—the electrical double layer—is formed. Its capacity is enormous, since the distance between the surface of the electrode and the plane passing through the centres of the ions at the closest approach to the electrode surface is about 0.1 nm (cf. page 121). When the electrical potential difference between the electrodes is changed, the voltage at this condenser is also changed and a certain charge must be supplied to (or taken from) the electrode. The abrupt increase of current followed by a rapid drop (see Figure 40a) corresponds to the charging of the electrode. This kind of current is termed the *condenser* or *charging current*. At a later stage the current is connected with the first of the main properties of the electrode since it is caused by the electrode reaction (a process of electron transfer between the electrode and the solution):

$$Fe^{2+} \underset{k_{red}}{\overset{k_{ox}}{\rightleftharpoons}} Fe^{3+} + e \tag{1}$$

The electrode reaction may be classified as one of heterogeneous reactions, several of which are known in chemistry. They occur at the boundary of two chemically different media (phases), in the present case of the metal of the electrode and of the electrolyte solution. Other heterogeneous reactions are processes of heterogeneous catalysis where, through the action of the catalyst surface, reactions take place that otherwise could not proceed, dissolving and crystallization, extraction from aqueous solutions by organic solvents, etc. The amount of substance reacting on a unit area of the boundary (interface) per second is called the rate of the heterogeneous reaction. For electrode reactions the participation of the charge from the electrode is significant. The reactivity of this charge depends on the electrical potential difference between the electrode and the solution. The electrode reaction rate is determined by the difference of the rate of oxidation (reaction (1) proceeding from the left to the right) and by the rate of reduction (for the opposite direction of (1)):

Electrode reaction rate = oxidation rate − reduction rate

$$v = k_{ox}c_{Fe^{2+}} - k_{red}c_{Fe^{3+}} \tag{2}$$

In this equation k_{ox} is the rate constant of the oxidation reaction, k_{red} the rate constant of the reduction reaction and $c_{Fe^{2+}}$ and $c_{Fe^{3+}}$ the concentrations of divalent and trivalent iron ions. The constants k_{ox} and k_{red} depend on the potential difference between the electrode and the electrolyte solution (the characteristics of this dependence will be discussed subsequently on pages 99 and 102). The current density corresponding to the electrode reaction rate (2) is

$$j = Fv \tag{3}$$

In such a relatively simple manner the electrode reaction occurs in the initial stage only. In the course of the oxidation reaction, the ions of ferrous iron are depleted in the surroundings of the electrode while the amount of ferric iron increases. Their concentrations at the electrode surface change compared with their values at a larger distance from the electrodes. The concentration gradients formed in this way result in diffusion transport of ferrous ions to the electrode and of ferric ions from the electrode. These ions could also be transported in the same direction by migration (in our experimental arrangement, however, the contribution of migration to the total transport process is small because of the presence of the excess indifferent electrolyte, $HClO_4$). The transport cannot, in fact, balance the concentration changes at the electrode completely; therefore reaction (1) slows down.

When the external voltage is switched off, first an instantaneous decrease by 1 mV in the potential difference between the electrodes is observed. This sudden decrease in the voltage is due to the disappearance of the *ohmic drop* in the potential in the solution. The electrolytic cell acts as an electrical resistance R, and when an electric current I flows through it, a part equal to IR of the voltage imposed on the electrodes is lost, mainly owing to the resistance of the electrolyte. The subsequent slow drop of the electrical potential difference is caused by the diffusion of ferric ions from the electrode and of ferrous ions to the electrode. Obviously, the potential difference in a currentless state depends on the concentrations of both ferric and ferrous ions, in general on the concentrations of the oxidant and of the reductant form of the species reacting at the electrode.

The reason for the electron transfer from the ferrous ion to the electrode can be found by consideration of the band structure of the electrode and similar characteristics of the oxidant and of the reductant in the solution. The average value of the energy of an electron in the solid phase is fixed by the energy of the Fermi level ε_F (cf. page 12) or, electrochemically speaking, by the electrochemical potential of the electron in the metallic phase $\tilde{\mu}_e$. (The energy of the Fermi level is usually referred to a single electron, whereas the electrochemical potential corresponds to a unit amount (1 mol) of electrons.) An attempt will now be made to find a relationship for $\tilde{\mu}_e$ in the solution. The oxidant Fe^{3+} and the reductant Fe^{2+} which are present do not peacefully coexist, but unceasingly react together. Electrons are transferred from the particles of the reductant to those of the oxidant so that the original Fe^{3+} becomes Fe^{2+} and vice versa. Chemically this exchange process is expressed by the equations

$$Fe^{2+} \longrightarrow Fe^{3+} + e$$
$$Fe^{3+} + e \longrightarrow Fe^{2+} \tag{4}$$

Thus, in the solution only a more or less fictitious concentration of electrons is present which corresponds to the electrons individually existing during the jumps from Fe^{2+} to Fe^{3+}. The situation is analogous to the existence of free protons in water. This minute concentration of electrons in the solution is determined by the highest energy level of the electron in the Fe^{2+} ion (also called the highest donor level or term), by the lowest non-occupied energy level of the electron in the Fe^{3+} ion (by the lowest acceptor term) and by the number of these levels that are available in a unit volume of the solution, i.e. by the concentration (for dilute solutions) of the Fe^{2+} and Fe^{3+} ions. Consequently, the Fermi level of the electron is situated between the highest occupied energy level of the electron in the reductant form (Fe^{2+}) and the lowest non-occupied energy level in the oxidant form (Fe^{3+}) of an oxidation–reduction system (see Figure 41).

With the help of an elementary rule of thermodynamics a simple way to calculate $\tilde{\mu}_e$ will now be outlined. Let us take a chemical reaction and substitute electrochemical potentials for the symbols of reactants: a relationship describing the equilibrium situation will be readily found. In the case of reaction (4),

$$\tilde{\mu}_{Fe^{2+}} = \tilde{\mu}_{Fe^{3+}} + \tilde{\mu}_e(\text{solution})$$

which gives (cf. equation (3), page 69)

$$\tilde{\mu}_e(\text{solution}) = \mu_{Fe}^0 - \mu_{Fe^{3+}}^0 + \frac{RT}{F}\ln\left(\frac{c_{Fe^{2+}}}{c_{Fe^{3+}}}\right) - F\varphi(\text{solution})$$

An equilibrium of two phases in contact is characterized by the condition that the electrochemical potential of a particle in one phase is equal to the electrochemical potential of the same particle in the other phase (when the particle is, in fact, present in this phase). For the electrochemical potential of

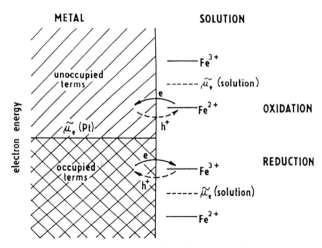

Figure 41. Relative position of occupied and unoccupied energy levels in a metal electrode and in the oxidized and the reduced forms in solution for oxidation and reduction. The relative position of the Fermi levels (expressed as electrochemical potentials of electrons $\tilde{\mu}_e$) decides whether an electron is transferred from the solution into the electrode (or a hole from the electrode into the solution) or in the opposite direction. (According to H. Gerischer)

each of the platinum electrodes in equilibrium it holds that

$$\tilde{\mu}_e(Pt) = \tilde{\mu}_e(solution) \tag{5}$$

The electrochemical potential of the electron in metal M is given by the equation

$$\tilde{\mu}_e(M) = \mu_e^0(M) - F\varphi(M)$$

This equation contains no term indicating a concentration or an activity of an electron in the solid phase, since this is constant and is included in the term $\mu_e^0(M)$. The change in the electron concentration could be accomplished by charging the metal, but the concentration of excess electrons (or holes) is negligibly small in comparison with their overall concentration.

Thus, for the electrical potential difference between the metal and the solution (the Galvani potential difference) we have*

$$\Delta_s^{Pt}\varphi \equiv \varphi(Pt) - \varphi(solution)$$

$$= \mu_{Fe^{3+}}^0 - \mu_{Fe^{2+}}^0 - \mu_e^0(Pt) + \frac{RT}{F} \ln \frac{c_{Fe^{3+}}}{c_{Fe^{2+}}}$$

Since the same equation holds for both of the electrodes of the electrolytic cell an electronic voltmeter indicates no potential difference between the two platinum electrodes. However, the remaining potential difference measured between the electrodes 1 and 2 after electrolysis is due to the change of the ratio of the *potential-determining species*, $c_{Fe^{3+}}/c_{Fe^{2+}}$, at the small electrode. Thus the potential difference of the electrode and solution differs between the two electrodes, and the measured voltage in the current-free state is

$$\Delta V = \Delta_s^{Pt}\varphi(2) - \Delta_s^{Pt}\varphi(1) = \frac{RT}{F} \ln \frac{c_{Fe^{3+}}(2)c_{Fe^{2+}}(1)}{c_{Fe^{3+}}(1)c_{Fe^{2+}}(2)}$$

The system consisting of two electrodes present in solutions differing only in the concentration of the potential-determining substances is called the *concentration cell*, which is the simplest member of the family of *galvanic cells* belonging to the more numerous group of *electrochemical cells*. The quantity ΔV is called e.m.f. (which stands for electromotive force but, since it is no force and the attribute 'electromotive' is rather cryptic, the anagram is preferred in denoting this concept).†

During electrolysis the system is out of equilibrium, which is displayed in the values of $\tilde{\mu}_e(Pt)$ of the small electrode. When it is larger than $\tilde{\mu}_e$ (solution) the

---

* In general, these electrical potential differences between the phases α and β, $\varphi(\alpha) - \varphi(\beta)$, are denoted by the symbol $\Delta_\beta^\alpha\varphi$.
† As already shown by J. W. Gibbs at the end of the last century, pure thermodynamics entitles us to measure electric potential differences exclusively between two chemically identical phases (metals, semiconductors and, with some reservation, also electrolyte solutions). Thus, the e.m.f. is the electric potential difference between electrodes (if their material is identical) or between chemically identical connections to the electrodes.

electrons prefer to jump from the electrode to the ferric ions. The small electrode then acts as the *cathode*. In the opposite case it becomes an *anode* (see Figure 41).

OXIDANTS AND REDUCTANTS

The strengths of various oxidants and reductants are different. Thus, for instance, the ion of trivalent iron, Fe^{3+}, oxidizes the ion of divalent vanadium, V^{2+}, according to the equation

$$Fe^{3+} + V^{2+} = Fe^{2+} + V^{3+}$$

whereas it is unable to oxidize the ion of monovalent thallium, Tl^+, to trivalent thallium, Tl^{3+}. This is, however, easily oxidized by the ion of tetravalent cerium, Ce^{4+}:

$$2\,Ce^{4+} + Tl^+ = 2\,Ce^{3+} + Tl^{3+}$$

Tetravalent cerium will also oxidize divalent vanadium. On the other hand, trivalent vanadium cannot oxidize the ferrous ion. Trivalent thallium oxidizes divalent iron but will not react with trivalent cerium. Obviously, the ability of various oxidants and reductants is due to the energies of their lowest acceptor (non-occupied) levels or their highest donor (occupied) levels. Therefore, oxidations and reductions of various substances at electrodes will also require appropriate values of the Fermi levels. It could be suggested that the energy of the Fermi level at an electrode immersed in a solution containing the investigated oxidation–reduction system should be measured. Unfortunately, this is not that simple. A direct method for determination of the Fermi level of an isolated single electrode is not possible because, for such a measurement, two electrodes are necessary. When the experimental arrangement described on page 76 is used, then, at equilibrium, an electrical potential difference equal to zero would always be obtained because the electrodes are made of the same material and are placed in identical solutions.

Obviously our system requires a radical change. For electrode 1 an electrode is chosen at which another reaction takes place than at electrode 2. The different reaction taking place at electrode 2 influences the corresponding Fermi level which will then be compared with that under the influence of the reaction occurring at electrode 1.

Oxidizing and reducing abilities of these reactants will then be quantitatively evaluated by means of *electrode potentials* (not to be confused with the potential difference of the electrode and solution). The two electrodes, together with suitable electrolyte solutions, form a galvanic cell. To mix two oxidation–reduction (redox) systems in the same solution with two submerged platinum electrodes would not be a good approach as they would react and, at equilibrium, the measured e.m.f. would be zero. Therefore, the solutions containing the redox systems are placed in separate vessels. These vessels are connected by a liquid bridge (p. 71). The resulting system is described as

$$
\begin{array}{c|c|c|c|c}
1 & 2 & 3 & 4 & 5 \\
\text{Pt} & \text{Fe}^{3+}, \text{Fe}^{2+} & \text{saturated KCl} & \text{V}^{3+}, \text{V}^{2+} & \text{Pt}
\end{array}
$$

The anions of the salts are not indicated and the concentrations are 10^{-3} M Fe^{3+}, 10^{-2} M Fe^{2+}, 10^{-1} M V^{3+} and 10^{-3} M V^{2+}. The vertical lines denote the boundaries between individual phases.

The purpose of the liquid bridge is twofold:

1. It prevents the solutions 2 and 4 from mixing (and, therefore, from reacting with each other).
2. It practically brings both solutions 2 and 4 to the same electric potential by removing the diffusion (liquid junction) potential.

Because of the latter property, the e.m.f. is obtained by subtraction of the electrical potential of the left-hand electrode from that of the right-hand electrode (this is mere convention prescribed by IUPAC for galvanic cell schemes of the type shown above).

Thus, the result obtained with the help of the equation for $\Delta_s^{Pt}\varphi$ on page 80 is

$$\Delta V = \Delta_s^{Pt}\varphi(\text{right-hand side}) - \Delta_s^{Pt}\varphi(\text{left-hand side})$$

$$= \mu_{V^{3+}}^0 - \mu_{V^{2+}}^0 - \mu_{Fe^{3+}}^0 + \mu_{Fe^{2+}}^0 + \frac{RT}{F} \ln \frac{c_{V^{3+}} c_{Fe^{2+}}}{c_{V^{2+}} c_{Fe^{3+}}} \tag{1}$$

At 25 °C the experimental value of ΔV is -0.77 V. The negative value of ΔV means that the system Fe^{3+}/Fe^{2+} is a stronger oxidant than the system V^{3+}/V^{2+}, but quantitative characteristics of this property attributed to individual redox systems are still needed. It is necessary, in the first place, to standardize the combinations of the standard chemical potentials, μ^0 values, pertaining to each redox system. This is achieved by choosing a definite redox system as a standard.

References

1. Page xiii, Ref. 2, Chapter 3.
2. H. Reiss, 'The Fermi level and the redox potential', *J. Phys. Chem.*, **84**, 3783 (1985).

ELECTRODE POTENTIALS

In aqueous solutions, which are of major concern in electrochemistry, hydrogen ions are always present. Therefore it is advantageous to use an electrode as a reference where a reaction occurs involving the participation of hydrogen ions. There are quite a few reactions of this type, but the simplest of them is

$$\tfrac{1}{2}H_2 + H_2O = H_3O^+ + e$$

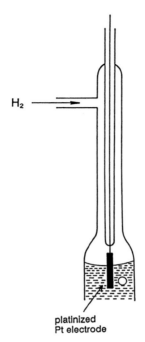

H₂

platinized
Pt electrode

Figure 42. A hydrogen electrode

Here gaseous hydrogen is the reductant and the oxonium ion is the oxidant.

In order to obtain a suitable hydrogen electrode, platinum or another platinum group metal is used as the electrode material; for alkaline solutions porous nickel can also be used. For good functioning the platinum electrode must be covered by a thin layer of deposited platinum sponge called 'platinum black' (cf. page 86). A hydrogen electrode is shown in Figure 42. At the hydrogen electrode gaseous hydrogen is oxidized to hydronium ions and, at the same time, oxonium ions are reduced to hydrogen, according to the above equation.*

The galvanic cell contains the Fe^{3+}, Fe^{2+} electrode as the redox system under investigation and the hydrogen electrode as a *reference electrode* is then described by the scheme

$$Pt|H_2|H_3O^+|\text{saturated } KCl|Fe^{3+}, Fe^{2+}|Pt$$

(again the anions have been left out). The platinum electrode is, of course, in

---

* The mechanism of this process is not simple and the final equilibrium between hydrogen, hydronium ions and the electrode consists of a number of successive equilibrium steps. Gaseous hydrogen is in equilibrium with hydrogen dissolved in water. The dissolved hydrogen adsorbs on the platinum surface with formation of atomic hydrogen (for more details, see page 109). The atomic hydrogen participates in the equilibrium with hydrogen ions and the final product, oxonium ions.

contact with the H_3O^+-containing solution but, at the same time, it comes into contact with gaseous hydrogen which is bubbled through the solution.

To calculate the potential difference of the electrode and solution for the hydrogen electrode the chemical potential of the gas H_2 must be considered. The pressure is usually a characteristic quantity of a gas instead of molar concentration (cf. page 32). Thus

$$\mu_{H_2} = \mu_{H_2}^0 + RT \ln \frac{p_{H_2}}{p_{H_2}^0}$$

where p_{H_2} is the pressure of hydrogen and $p_{H_2}^0$ is the standard pressure, 10^5 pascals. The chemical potential of the solvent, water, is equal to the standard chemical potential as, except at very high concentrations, the ions only exhibit a negligible influence on the bulk solvent. Thus, the potential difference hydrogen electrode/solution is

$$\Delta_s^{Pt} \varphi = \frac{\mu_{H_3O^+}^0 - \frac{1}{2}\mu_{H_2}^0 - \mu_{H_2O}^0}{F} + \frac{RT}{F} \ln c_{H_3O^+} - \frac{RT}{2F} \ln \frac{p_{H_2}}{p_{H_2}^0}$$

The hydrogen electrode is in its standard form when $c_{H_3O^+} = 1$ (that is, $a_{H_3O^+} = 1$) and $p_{H_2} = p_{H_2}^0$. The principal convention is that the combination of standard chemical potentials, $\mu_{H_3O^+}^0 - \frac{1}{2}\mu_{H_2}^0 - \mu_{H_2O}^0$, is equal to zero, which implies that the potential difference of the standard hydrogen electrode and solution is also equal to zero and the e.m.f. of the cell on page 83 is given by the relationship

$$\Delta V \equiv E(Fe^{3+} + e = Fe^{2+})$$

$$= \frac{\mu_{Fe^{3+}}^0 - \mu_{Fe^{2+}}^0}{F} + \frac{RT}{F} \ln \frac{c_{Fe^{3+}}}{c_{Fe^{2+}}}$$

$$= E^0(Fe^{3+} + e = Fe^{2+}) + \frac{RT}{F} \ln \frac{c_{Fe^{3+}}}{c_{Fe^{2+}}}$$

The term $E(Fe^{3+} + e = Fe^{2+})$ or, abbreviated, $E_{Fe^{3+}, Fe^{2+}}$ is the *electrode potential* (redox potential) while $E^0(Fe^{3+} + e = Fe^{2+})$ or $E_{Fe^{3+}, Fe^{2+}}^0$ is the *standard electrode potential* versus the standard hydrogen electrode (SHE) of the redox system of trivalent and divalent iron (remember that these are not electric potential differences of the electrode and solution but experimentally determined e.m.f.'s).

In the case of the redox reaction, ox + z e = red, the above equation acquires the form

$$E(ox + z\,e = red) = E^0(ox + z\,e = red) + \frac{RT}{zF} \ln \frac{c_{ox}}{c_{red}}$$

which is the *Nernst equation*.

Since these values are standardized with respect to the standard hydrogen electrode (SHE) they are called 'on the hydrogen scale'. For various inorganic systems these values have been set up in Figure 43.

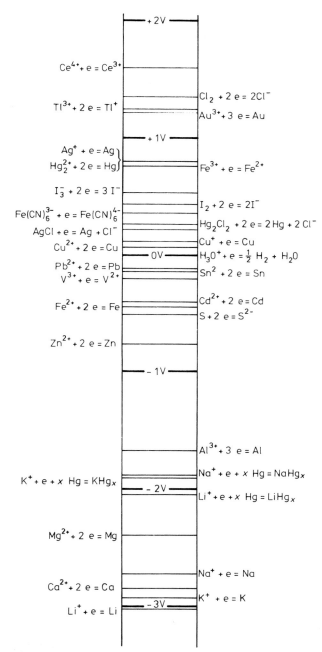

Figure 43. Standard potential scale of inorganic systems

In the same way the e.m.f. of the cell on page 82 will be discussed. Its e.m.f. is given by the simple equation

$$\Delta V = E_{V^{3+},V^{2+}} - E_{Fe^{3+},Fe^{2+}}$$

or, in general (cf. page 82),

$$\Delta V = E(\text{right-hand electrode}) - E(\text{left-hand electrode}) \tag{1}$$

The standard electrode potential characterizes the oxidizing or reducing ability of the component of oxidation–reduction systems. With a more positive standard electrode potential, the oxidized form of the system is a stronger oxidant and the reduced form is a weaker reductant. Similarly, with a more negative standard potential, the reduced component of the oxidation–reduction system is a stronger reductant and the oxidized form a weaker oxidant.

The values of standard electrode potentials help to explain the course or reactions (page 81). The standard electrode potential of the Fe^{3+}/Fe^{2+} system is $E^0_{Fe^{3+},Fe^{2+}} = 0.771$ V, while for the V^{3+}/V^{2+} system it is $E^0_{V^{3+},V^{2+}} = -0.20$ V. The Tl^{3+}/Tl^+ system is characterized by $E^0_{Tl^{3+},Tl^+} = 1.25$ V so that univalent thallium is not oxidized by trivalent iron but by tetravalent cerium ($E^0_{Ce^{4+},Ce^{3+}} = 1.610$ V). It is important to note that a reaction will not occur in practice unless it has sufficient velocity. This condition is fulfilled with the majority of the examples of this paragraph while, for example, the direct oxidation of hydrogen by trivalent iron does not take place in water.

In the above examples the course of a reaction is completely determined by standard potential values because their differences are large. When the standard potentials are not very different the direction of the reaction depends on the values of the electrode potentials given by the complete Nernst equation. Equilibrium is attained in the oxidation–reduction reaction when both reacting systems have the same redox potential.

The electrode potential is not determined exclusively by the two components of the oxidation–reduction system present in the solution. A different case has already been mentioned for the hydrogen electrode, which is an example of a gas electrode (the reduced form is gaseous hydrogen). A cationic electrode is an electrode consisting of metal M immersed in a solution containing ions of the same metal M^{z+}. The reaction at this electrode is

$$M^{z+} + z\,e = M$$

The Nernst equation for a cationic electrode is

$$E_{M^{z+}/M} = E^0_{M^{z+}/M} + \frac{RT}{zF} \ln c_{M^{z+}}$$

since the activity of the reduced form, i.e. of the electrode material, has been included as a constant term in $E^0_{M^{z+}/M}$.

It is not easy to obtain a perfectly functioning hydrogen electrode without observing several strict precautions: the hydrogen must be pure, the preparation of the electrode covered with platinum black is not simple and it is necessary to prevent various impurities from reaching the neighbourhood of the electrode since they would 'poison' the catalytically active layer. In view of this, *electrodes of the second kind* are almost exclusively used as reference electrodes. These are metallic electrodes covered with a sparingly soluble salt of the corresponding metal ion with a suitable anion; a soluble salt of this anion with an alkali metal cation is present in the solution in contact with the electrode. The layer of the sparingly soluble salt is simultaneously in equilibrium with the metal and with the electrolyte solution.

The silver–silver chloride electrode consists of a piece of silver wire or a plate covered with a layer of silver chloride and immersed in an electrolyte solution containing chloride ions (e.g. KCl).

Silver chloride is a substance with a very low solubility. The higher the chloride concentration in the solution the lower its solubility, which is quantitatively expressed by the expression for the solubility product

$$P_{AgCl} = c_{Ag^+} c_{Cl^-}$$

On inserting $c_{M^{z+}} = c_{Ag^+}$ from this equation into the equation for a cationic electrode the Nernst equation for the silver–silver chloride electrode becomes

$$E_{AgCl} = E(AgCl + e = Ag + Cl^-)$$

$$= E^0_{AgCl/Ag} - \frac{RT}{F} \ln c_{Cl^-}$$

$$= 0.222 - 0.0591 \log c_{Cl^-} \quad \text{(volts)} \quad \text{(at 25 °C)}$$

An analogous equation is valid for the calomel electrode which consists of a mercury electrode covered by calomel (HgCl) and a potassium chloride solution. The saturated calomel electrode containing a saturated solution of potassium chloride is particularly important ($E_{HgCl/Hg}$(sat. KCl) = 0.242 V versus SHE).

These electrodes are also used as electric potential probes in electrolytes where different points possess different values of potential. The electric potential difference between such electrodes indicates the electric potential difference between these points, for example, during electric current flow.

This approach has found frequent application in electrophysiology. In order to determine the electric potential difference, for example between two points in the extracellular liquid or between the inside and the outside of a cell (the membrane potential, see page 139), micropipettes (Figure 44) are often used. These are silver–silver chloride electrodes filled with a saturated KCl solution immobilized with agar-agar gel. A capillary with a tiny tip (diameter 1–3 μm) connects the electrode to the point of measurement. Since the potassium ions often influence the electric properties of excitable cells (see page 173), the saturated KCl solution is sometimes replaced with saline (0.9 % NaCl solution). In this case, of course, the liquid junction potential between the orifice of the capillary and the investigated solution cannot be neglected. On the other hand, a probe of this type correctly reflects the changes in electric potential.

88

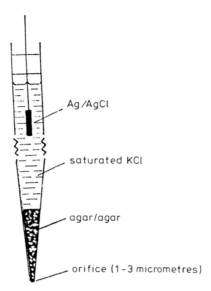

Figure 44. A micropipette

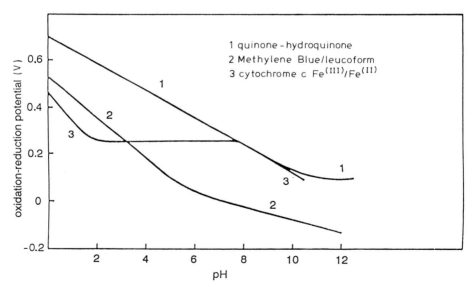

Figure 45. The pH dependence of organic oxidation–reduction potentials. The slope dE/dpH of the dependence for the quinone–hydroquinone system is -0.059 V below pH 10, then -0.029 V and 0 V. The slope for the methylene blue–leucoform of methylene blue is -0.089 V for pH < 6 and then becomes -0.029 V. The slope for the ferriferro-cytochrome c is -0.118 V below pH 2, 0 V between pH 2 and 8, and -0.059 V at higher pH. The change from the oxidized form to the reduced form of this enzyme does not affect the acid–base properties of its protein component

More complicated electrode equilibria are found in organic oxidation–reduction systems which are often of quinoid character. The equilibrium between a quinone-type oxidized form and the hydroquinone-type reduced form is described by the equation

$$\text{Quinone} + 2\,e = \text{hydroquinone}$$

Both forms, or at least the reduced form, are acids binding different numbers of hydrogen ions. Such an oxidation–reduction system is often characterized by the pH dependence of the redox potential, E_H^0, defined by the condition that the total concentration of all the components in the oxidized form (protonized to different degrees) is equal to the total concentration of all the components in the reduced form. Examples of this dependence are shown in Figure 45. Curve 1 corresponds to the benzoquinone–benzohydroquinone system. In an aqueous solution the quinone possesses neither acidic nor basic properties, while the hydroquinone is a diprotonic acid. Curve 2 shows the pH dependence of the redox potential of methylene blue. The acidobasic properties of this oxidation–reduction system consisting of a blue oxidized form and of a colourless reduced form (called the leucoform from Greek *leukos* = white) are demonstrated in the following reaction scheme:

pH < 6:

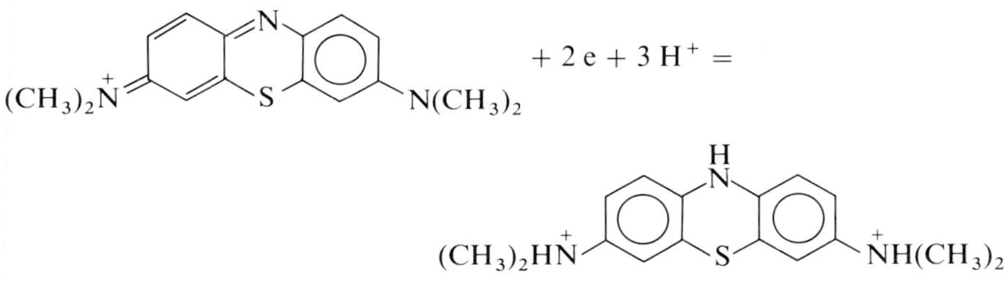

$$+ 2\,e + 3\,H^+ =$$

pH > 6:

$$+ 2\,e + H^+ =$$

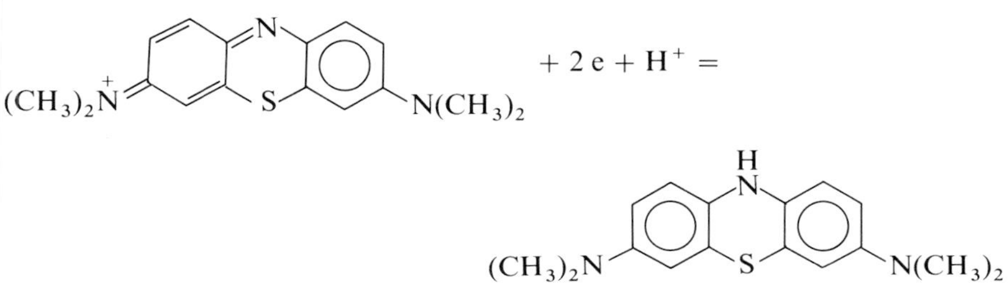

Cytochrome c, an enzyme (Figure 46), contains a porphyrin complex of iron as the prosthetic group with the oxidation–reduction properties described schematically by the Nernst equation (page 84). The protein component of the enzyme is a polyelectrolyte. The protolytic reactions of the prosthetic group lead to the pH dependence of the redox potential (curve 3, Figure 45).

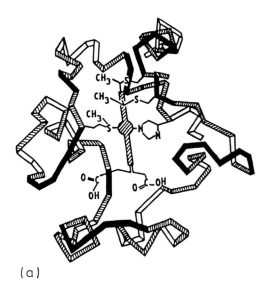

(a)

(b)

Figure 46. Cytochrome c structure according to Dickerson and coworkers. In scheme (a) the individual amino acids are represented by trapezoids. Shading indicates the distance from the observer. The rod in the centre is the lateral view of the heme, the porphyrin complex of iron. The sulphur atom of methionine and the nitrogen atom of histidine are located in the two remaining free coordination sites of the heme. The bounding of the protein chain to the upper part of the heme is shown in scheme (b)

Transformation of quinone to hydroquinone is, in fact, a combination of two reactions of successive binding of two electrons to the quinone molecule. First, the semiquinone is produced from the quinone:

$$\text{Quinone} + e = \text{semiquinone}$$

and then the hydroquinone is formed:

$$\text{Semiquinone} + e = \text{hydroquinone}$$

The semiquinone is often quite unstable so that its formation cannot be detected. The reduction of quinone is then a simple two-electron reaction. Sometimes, however, the semiquinone is relatively stable in spite of the fact that it is a free radical (it contains one unpaired electron).

References

1. M. S. Antelman and F. J. Harris, *The Encyclopedia of Chemical Electrode Potentials*, Plenum Press, New York, 1982.
2. A. J. Bard, J. Jordan and R. Parsons (eds.), *Oxidation–Reduction Potentials in Aqueous Solution*, Blackwell, Oxford, 1986.
3. W. M. Latimer, *Oxidation Potentials*, Prentice-Hall, New York, 1952.
4. S. Wawzonek, 'Potentiometry; oxidation reduction potentials', in *Techniques of Chemistry* (eds. A. Weissberger and B. W. Rossiter), Vol. I, Part IIa, Wiley-Interscience, New York, 1971.

ELECTROCHEMICAL CELLS: SOURCES AND SINKS OF ENERGY

An *electrochemical cell* consists of two (or more) electrodes submerged in an electrolyte solution (or in a molten electrolyte or in contact with a solid electrolyte). It may be in a current-free state (which may or may not actually be a true equilibrium) or supply electric energy (a *galvanic cell*) or accept electric energy from an external source (an *electrolytic cell*). The properties of an electrochemical cell will be demonstrated on the system shown by the scheme on page 82. The e.m.f. of this cell is given by equation (1) (page 86). If, for example, a little d.c. electromotor is connected to the electrodes, the electric current produced performs mechanical work. At the same time a *cell reaction*

$$V^{2+} + Fe^{3+} = V^{3+} + Fe^{2+}$$

proceeds by oxidation of V^{2+} to V^{3+} at the *negative electrode* (which now acts as an *anode*) and by reduction of Fe^{3+} to Fe^{2+} at the positive electrode (now a *cathode*). Inside the cell the positive current carried by ions flows from the anode to the cathode while in the external circuit it flows from the cathode to the anode (by definition in the opposite direction to the flow of electrons). Thus, the electrochemical cell functions as a *galvanic cell*, i.e. the *electrochemical source of current* where the chemical energy of the cell reaction is converted (transduced) to electric energy (which is subsequently converted to the mechanical energy of the electromotor).

Now connect the above electrochemical cell to a source of voltage which is larger than the e.m.f. of the cell (this means that the potential of the left-hand electrode becomes 'more positive' and that of the right-hand electrode 'more negative'). The current will flow through the circuit in the opposite direction. Therefore the cell reaction proceeds from the left to the right. Under these conditions the system acts as an *electrolytic cell*. Electric energy supplied to the cell is converted to chemical energy by increasing the V^{2+} concentration at the expense of V^{3+} concentration, while in the vicinity of the positive electrode the Fe^{3+} concentration grows and that of the Fe^{2+} diminishes. The ability of the present system to act both as a current source and as an electrolytic cell making use of the same cell reaction is called the *reversibility* of the cell.

The electric work supplied or gained by a reversible electrochemical cell is the maximum useful energy (the Gibbs energy) defined on page 68. It is calculated as follows. The above cell reaction (like any chemical reaction) consists of products of the amounts of substance (moles) of the reactants expressed by their chemical symbols and of the stoichiometric coefficients indicating the number of these units participating in the reaction (in the present case, all the stoichiometric coefficients are equal to unity). The change of the Gibbs energy ΔG corresponding to such an extent of the reaction is given by the difference of the sum of chemical potentials of the products of the reaction multiplied by the stoichiometric coefficients and of an analogous sum for the original reactants. Thus, in the present case,

$$\Delta G = \mu_{V^{3+}} + \mu_{Fe^{2+}} - \mu_{V^{2+}} + \mu_{V^{3+}}$$

The electric work gained from the transduction of this Gibbs energy change is $F\Delta V$, which has a negative value (cf. page 82). The negative value of ΔG indicates that the reaction proceeds spontaneously.

A general rule can now be formulated for energy transformations in a galvanic cell. The cell reaction has to be written in such a way that oxidation takes place at the right-hand electrode. If the reduced component of the redox system at the right-hand side is oxidized to the oxidized form spontaneously (by closing the external circuit) then the Gibbs free energy change given by the relationship $\Delta G = zF\Delta V$ as well as the e.m.f. is negative and, at the same time, z moles of electrons are conducted from the right-hand electrode through the external circuit to the left-hand electrode in order to bring about the cell reaction in the form it has been written. If the opposite process, the oxidation of the reductant, had to proceed at the left-hand electrode, then for a spontaneous process ΔG as well as ΔV should be positive. The *charge member z* of the cell reaction is unity only in the present case while, for example, for reactions like $2\,Fe^{3+} + Zn$ (metal) $= 2\,Fe^{2+} + Zn^{2+}$ it is equal to 2.

The objective of e.m.f. calculation is reached by both methods, either based on the difference of electrode potentials or on thermodynamic data, provided the reversibility of the cell is preserved.

It should be noted that if the electrochemical cell acts as a *source of current* the voltage supplied is always smaller than the equilibrium e.m.f. This is caused by two factors. First, the current must overcome the internal resistance of the cell resulting in the internal ohmic drop of potential. The other factors are connected with the rates of the electrode reactions and of the transport of the reactants to the electrode, which requires a definite increase of potential in order to proceed (for a discussion see the section on electrode reactions, particularly page 104). On the other hand, for an *electrolytic cell* the required voltage is larger than the equilibrium e.m.f. as both main factors appearing in the case of a galvanic cell are in full strength but act in the opposite direction. The difference between the voltage necessary for the current to flow and the equilibrium e.m.f. is called the *overpotential* (overvoltage or polarization), usually determined after correction for the ohmic potential drop.

Most of the practical galvanic cells do not fulfil the condition of reversibility. Because of the sluggishness of the electrode reactions the e.m.f. depends on partial reaction steps which are sometimes not even reversible (i.e. after current reversal another reaction takes place in the cell, etc.). For example, in the oxygen–hydrogen cell (Grove's cell)

$$Ni \mid H_2 \mid KOH, H_2O \mid O_2 \mid Ag$$

the expected cell reaction

$$O_2 + 2 H_2 = 2 H_2O$$

does not occur under the current drain with an e.m.f. of 1.23 V. The reversible hydrogen oxidation reaction occurs at the porous nickel electrode but at the right-hand electrode oxygen is partially reduced to the hydrogen peroxide anion:

$$O_2 + H_2O + 2 e = HO_2^- + OH^-$$

(cf. page 111). The hydrogen peroxide formed is then decomposed on the electrode by a set of reactions to water and oxygen. When the current is reversed water is not oxidized to oxygen but oxidation of the electrode takes place:

$$2 Ag + 2 OH^- = Ag_2O + H_2O$$

These processes result in an e.m.f. of less than 1 V, which is much less than expected from the cell reaction.

On the contrary, when the lead storage battery

$$Pb \mid PbSO_4 \mid H_2SO_4, H_2O \mid PbO_2 \mid Pb$$

is discharged (the external voltage is lower than the e.m.f.) the cell reaction

$$PbO_2 + Pb + 2 H_2SO_4 = 2 PbSO_4 + 2 H_2O$$

takes place. When the lead battery is charged, the same cell reaction commences in the opposite direction. This cell reaction corresponds to the overall chemical transformation in the cell during charging and discharging. Therefore, the lead accumulator is an example of a reversible galvanic cell.

The relationship between the e.m.f. and ΔG (page 92) determines the upper limit for the transformation of the energy of a chemical reaction to the useful (electrical) energy with maximum efficiency. The Gibbs energy change is the measure of the exploitability of the energy of a chemical reaction regardless of the possibility of achieving this degree of exploitation.

When producing electrical energy in thermal power plants we also employ a chemical reaction, namely the combustion of carbonaceous fuels like coal, oil or natural gas; for example,

$$C + O_2 = CO_2$$

This reaction is connected with the release of the reaction heat (negative reaction enthalpy) $- \Delta H = 393$ kJ. The steam formed by feeding this amount of heat to the boiler drives the turbine, cools down at the same time and, at last, supplies heat to a heat exchanger and condenses to water. The utilization of the gained energy 393 kJ per 1 mol of carbon is limited by the efficiency of the heat engine. The heat engines include the steam engine, the steam turbine, the internal combustion engines, the magnetohydrodynamic power generator, the thermionic energy exchanger, etc. These are all devices that transform heat to other energy forms, mechanical or electrical. The heat is gained from a thermal source with a definite temperature while part of it is supplied to a heat exchanger with a lower temperature. The efficiency of energy transformation η is given by the ratio of the gained useful work $- W$ and of the heat supplied from the thermal source $- Q$:

$$\eta = \frac{W}{Q} = \frac{T_1 - T_2}{T_1}$$

where T_1 is the temperature of the thermal source and T_2 is that of the cooler heat exchanger (in kelvins). For high-quality power stations, $\eta \approx 0.45$.

The galvanic cell is not a thermal engine. It directly converts the maximum share of the energy of a chemical reaction (i.e. the Gibbs energy) to useful, electrical work.

The Gibbs energy change $\Delta G < 0$ and the heat gained from a chemical reaction $\Delta H < 0$ are related by the equation

$$\Delta G = -\Delta H + T \Delta S$$

where ΔS is the entropy change during the reaction. ΔS may sometimes be positive so that the efficiency of the energy transformation

$$\eta = \frac{\Delta G}{\Delta H}$$

is larger than unity. As the product $T \Delta S$ is the heat supplied to the cell from its surroundings when the cell reaction takes place close to equilibrium, even this amount of heat is transformed to useful work. The lead accumulator is an example of this situation. The entropy change accompanying its cell reaction is $+42.6 \, \mathrm{J \, K^{-1} mol^{-1}}$, so that the heat from the surroundings also contributes somewhat to the transformation of the chemical energy to electrical energy when discharging the battery. This energy balance is in no way contrary to the second law of thermodynamics, for when charging the accumulator an equal amount of heat is transferred to the surroundings.

Reference
1. Page xiii, Ref. 2, Chapter 3.

PRACTICAL GALVANIC CELLS

From the standpoint of the origin of chemical energy, practical galvanic cells can be divided into three groups.

Primary cells already contain the reagents of their energy-producing cell reactions when they are assembled. The most frequently used member of this group is the Leclanché or manganese dioxide cell (e.m.f. = 1.55 V). It was invented by the French engineer Georges-Lionel Leclanché (1838–1882) in 1866. It contains a carbon electrode covered with a layer of a mixture of manganese dioxide (pyrolusite) and soot suspended in a solution of ammonium chloride. The other electrode is made of zinc, which is usually amalgamated. The supply of electric current is connected with the cell reaction

$$MnO_2 + Zn + 8 \, NH_4^+ + 2 \, H_2O = Mn(NH_3)_4^{2+} + Zn(NH_3)_4^{2+} + 4 \, H_3O^+$$

The usual version of the Leclanché cell, called the dry cell (Figure 47), was invented much later. The cell is not actually dry, however, since it contains a solution of ammonium chloride thickened with wheat flour. This cell represents

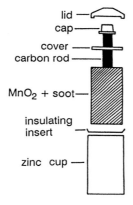

Figure 47. A dry cell

90% of global production of primary cells; production in the United States equals more than three billion cells per year.

Another important primary cell is the Weston cell:

$$\text{CdHg (12.5 wt \% Cd)} \mid \begin{array}{l} \text{3 } CdSO_4 \cdot 8 \text{ } H_2O \\ \text{saturated} \\ \text{aqueous} \\ \text{solution} \end{array} \mid \begin{array}{l} Hg_2SO_4 \\ \text{solid} \end{array} \mid Hg$$

with the e.m.f. $= 1.01807$ V at $25\,°C$. The Weston cell represents a standard of voltage because, when carefully prepared, it retains the required e.m.f. value with great accuracy. The temperature coefficient of the cell is small and a constant composition can be maintained for a long time.

Similar cells used as sources for miniature electrical appliances (hearing aids, pacemakers, etc.) is the mercury cell which is similar to the Weston cell with cadmium replaced by zinc or the indium cell with indium instead of cadmium. Superdry cells have a solid electrolyte, for example the cell

$$Ag \mid AgBr \mid CuBr \mid Cu$$

The solid electrolytes AgBr and CuBr have very low conductivity, so that the cell only acts as a long-life voltage source.

Secondary cells (accumulators, storage batteries) are galvanic cells where the power supplying reagents are formed only when the cell is charged from an external source of electric current. Thus, they serve for accumulation of electric energy. The most important (and the oldest) secondary cell is the lead accumulator. The originators of this storage battery are usually given as H. Sinsteden (1854) and G. Planté (1859), but it may be concluded from a much older Juvenile Lecture by M. Faraday that he was acquainted with the principle of the reaction in this cell. A virtually working lead accumulator was probably first constructed by Planté in 1860. Lead accumulators are used primarily as batteries in cars, as power sources for railway signals, for runabouts (the only *electromobiles*, electricity-driven motorcars, really used in practice), etc.

The charging/discharging twin process is called the accumulator cycle. A starter battery must last for 400 cycles. A disadvantage of the lead accumulator

is its large weight per unit of stored charge. Better weight parameters are exhibited by nickel–cadmium and silver–zinc batteries which are used in portable appliances (e.g. radios, television sets) and aircraft.

Among the new types of storage batteries the sodium–sulphur cell is much publicized. One of its electrodes consists of graphite immersed in molten sodium polysulphide (Na_2S_x) and the other of sodium amalgam. A solid–electrolyte wafer or tube (made of β-alumina, see page 18) separates these electrode systems. This cell could provide a power source for an electromobile to be used in city transport.

For other storage batteries lithium amalgam has been suggested as the negative electrode, together with an electronically conducting polymer (see page 20).

The third type of galvanic cells comprises *fuel cells*. Here, the cell reagents are continuously supplied to the cell during current drain. An ancient dream of electrochemists was the realization of the reaction of fuel burning as a practicable cell reaction. This would mean that coal would be oxidized at the negative electrode of the cell while oxygen would be reduced at the positive electrode. An attempt was made to construct such a cell, for example by the German chemists Wilhelm Ostwald and Fritz Haber (both Nobel Prize winners, but not for fuel cell discovery). At elevated temperatures, direct coal oxidation in a cell is possible, particularly when a fused electrolyte is used, but such cells are unstable and unsuitable for a long operation. Somewhat better results were achieved using carbon monoxide as a cell fuel. Oxidation occurs in a strongly acidic solution with platinum or palladium anodes. In another version a solid electrolyte, the Nernst mass ($85\% \, ZrO_2 + 15\% \, Y_2O_3$), is used and carbon monoxide is oxidized and oxygen reduced at platinum electrodes.

Compounds like methanol, formic acid, oxalic acid, etc., show better performance as fuels. At higher temperatures in strongly acidic solution even simple hydrocarbons like methane, ethane, and ethylene are also oxidized. However, none of these systems has been used in practical applications.

The only exception is the oxygen–hydrogen cell mentioned on page 93 which also belongs among fuel cells because gaseous hydrogen is the fuel. This cell does not transform chemical energy into electrical energy ideally but has the advantage of low weight per amount of energy supplied. This low weight was crucial in employing the oxygen–hydrogen fuel cells as the electric energy sources in space vehicles. The American space ship *Gemini* (Latin *gemini* = twins, as it carried two astronauts on board) contained this cell equipped with platinum electrodes and an ion-exchanging membrane (see page 139) as the cell electrolyte (a thin film of liquid electrolyte had to be present at the surface of each electrode contacting the membrane). The membrane simultaneously prevents the reacting gases from reaching the electrodes which they do not correspond to, i.e. the hydrogen to the oxygen electrode and vice versa. In the subsequent Apollo missions (which culminated in the lunar landing of the astronauts in 1969) oxygen–hydrogen cells with a liquid electrolyte were used (Figure 48).

98

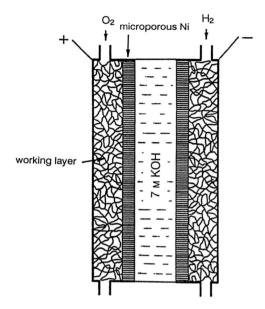

O_2 microporous Ni H_2

7 M KOH

working layer

Figure 48. A hydrogen–oxygen fuel cell with liquid electrolyte. The pores of the microporous nickel layer are filled with the electrolyte solution and prevent the gases from entering the electrolyte. The working layer consists of a nickel matrix with embedded patches of Raney nickel (hydrogen electrode) or dispersed silver (oxygen electrode). In the pores of the working layer a three-phase (metal/electrolyte/gas) boundary is formed. The electrode reaction occurs close to this boundary

In addition to chemical energy, light energy can be transformed into electrical energy in electrochemical devices. These electrochemical photovoltaic cells will be dealt with in the section on photoelectrochemistry (page 132).

References

1. A. J. Appleby, F. R. Foulkes and V. N. Reinhold, *Fuel Cell Handbook*, Reinhold, New York, 1989.
2. V. S. Bagotzky and A. M. Skundin, *Chemical Power Sources*, Academic Press, New York, 1980.
3. M. W. Breiter, *Electrochemical Processes in Fuel Cells*, Springer, New York, 1969.
4. B. D. McNichol and D. A. J. Rand (eds.), *Power Sources for Electric Vehicles*, Elsevier, New York, 1984.
5. W. Vielstich, *Fuel Cells: Modern Processes for the Electrochemical Production of Energy*, Wiley-Interscience, New York, 1970.
6. C. A. Vincent, A. F. Bonino, M. Lazari and B. Scrosati, *Modern Batteries, An Introduction to Electrochemical Power Sources*, Edward Arnold, London, 1984.

THEORY OF ELECTROLYSIS

Let us return to the experiment described on page 76. Here the dependence of the electric current on the electrode potential during current flow (the polarization curve or the voltammogram) is of interest. The experimental arrangement shown in Figure 39 has, however, a definite drawback. The ohmic potential drop formed between electrodes 1 and 2 (see page 78) cannot usually be neglected. Thus, in order to record the required dependence, it must be eliminated in a suitable way. When the resistance of the whole electrolysis system R (i.e. the resistance between the metallic leads to the electrodes) is known, then the quantity IR can be simply subtracted from the voltage imposed on the electrodes. As a rule, however, a three-electrode arrangement* is used (see Figure 49). The reference electrode (either made of the same material and present in the same solution as the polarized electrode or a calomel or silver–silver chloride electrode with a salt bridge) is attached through the thin glass tube (the Luggin capillary) as close as possible to the polarized (working or indicator) electrode. Under the flow of electric current between the indicator (polarized) electrode 2 and electrode 1 (called the auxiliary or current-supplying electrode) the difference in electrical potential between the indicator and the reference electrode, $E - R_{ref}$, the *polarization* of the electrode, is measured potentiometrically. Since the potential of the reference electrode E_{ref} is known, this measurement yields the polarization curve. The whole procedure is very frequently carried out automatically using an electronic or computer-controlled potentiostat. This device maintains continuous control of the voltage applied between the indicator and the reference electrode. This voltage, fed from a suitable source, may be constant or may follow a required time-dependence

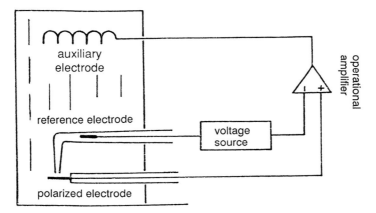

Figure 49. A simplified scheme of a system for potentiostatic polarization of the electrode

---

* In polarography a two-electrode system is often used since the IR drop can be neglected in view of rather low currents.

programme (linear increase or decrease, triangular function, step-wise potential change, etc.).

A potentiostat is not essential for the investigation of rather slow electrode reactions where the composition of the electrolysed solution is not changed even in the closest vicinity of the electrode. A suitable example is the reduction of the hydrogen ion in solution of rather concentrated hydrochloric acid. The resulting polarization curve is close to an exponential (Figure 50a).

In the case of rapid electrode reactions the concentrations of the reacting species undergo a change in the neighbourhood of the electrode so that the course of electrolysis is under the influence of the transport of reactants to the electrode and of the products in the opposite direction. To obtain reproducible voltammograms either the transport alone has to be controlled by convection or by special arrangements of the system, for example by using dropping mercury electrode in polarography or an ultramicroelectrode (see pages 127 and 131), or the time dependence of the current has to be measured with careful control of the overpotential using a potentiostat.

A survey of various types of voltammograms is shown in Figure 50. If a rotating or vibrating electrode is used the transport process is a steady-state convective diffusion. The system from page 76 will be modified in such a way that only ferric ions are present in the solution. The voltammogram is a sigmoid curve (a so-called wave, see Figure 50c). At the beginning the current is controlled by the rate of the electrode reaction but with increasing positivity of the electrode potential the concentration drop at the electrode becomes more pronounced. Finally, the concentration of Fe^{2+} close to the electrode approaches zero and a further increase of potential does not result in a further increase of current so that a horizontal 'plateau' is formed on the voltammogram. This current, called the *limiting current*, is a quantity of considerable importance for analysis of solutions as it is directly proportional to the concentration of the electroactive substance. The position of the wave on the potential coordinate is characterized by the *half-wave potential*, $E_{1/2}$, which is the electrode potential at which the current has reached a half of the limiting current value.

If the solution is not stirred the electrode can be polarized in several different ways. A rectangular voltage pulse can be applied to the electrode and the change (usually decay) of current recorded as a function of time (a *chronoamperometric curve*). A *peaked voltammogram* is obtained by application of a potential pulse which grows linearly with time. Another version is a triangular pulse (Figure 50b) where, for example, in one branch the electroactive substance is reduced and in the other branch the product is reoxidized (this is not always the case). This procedure can be repeated as long as a steady picture, the *cyclic voltammogram*, is obtained.

Peaks on the voltammograms originate from the following sequence of processes. When increasing the electrode potential a roughly exponential rise of current is observed as the concentration of the original electroactive substance at the surface of the electrode only slowly falls while the concentration of the

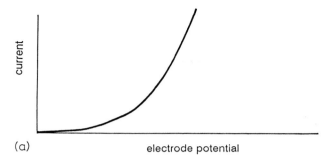

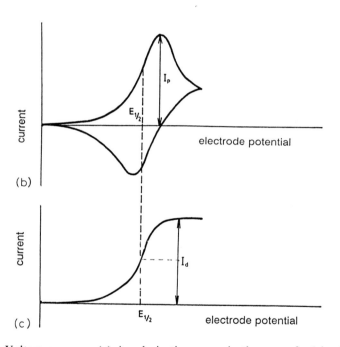

Figure 50. Voltammograms. (a) A polarization curve in the case of a 'slow' electrode reaction. The concentration of the electroactive substance is so high that its transport to the electrode does not influence the current (the electrode potential corresponds to the initial portion ('foot') of the other two voltammograms). (b) A typical voltammogram obtained when the electrode is polarized with a triangular voltage pulse (the electrode reaction is 'reversible', i.e. the product is converted back to the original species). The peak current I_p is directly proportional to the concentration of the electroactive substance. The half-wave potential defined in the last voltammogram is not situated in the middle of the rising part of the voltammogram. (c) The voltammogram obtained in the case of the steady-state transport to the electrode or of polarography. The limiting diffusion current I_d is directly proportional to the concentration of the electroactive substance. The electrode potential corresponding to the half of the limiting current value is the half-wave potential $E_{1/2}$

product increases. However, on further increasing the potential a marked exhaustion of the electroactive substance in the vicinity of the electrode ensues, which slows down the increase of the current and, finally, the peak is reached. When increasing the potential further the concentration of the electroactive substance at the surface of the electrode drops to zero and the decline of the current approaches the dependence described on page 60 by the equation for the dependence of the concentration gradient in linear diffusion. (Note that the potential is directly proportional to time; therefore the voltammogram is a dependence of current on potential as well as on time!) The peak current is again directly proportional to concentration of the electroactive substance and, in this way, can be exploited for analysis.

Reference

1. Page xiii, Ref. 1, Chapter 6, and Ref. 2, Chapter 5.
2. R. G. Compton and H. Hamnett (eds.), *New Techniques for the Study of Electrodes and Their Reactions*, Elsevier, Amsterdam, 1989.
3. P. Delahay, *New Instrumental Methods in Electrochemistry*, Interscience, New York, 1954.
4. Southampton Electrochemistry Group, *Instrumental Methods in Electrochemistry*, Ellis Horwood, Chichester, 1985.

ELECTRODE REACTIONS

The term *electrode reaction* was used on page 77 where the dependence of its rate on the electrode potential was pointed out. This was displayed in the changing values of the reduction (cathodic) and oxidation (anodic) electrode reaction constants. Otherwise no change of the current due to the change of the potential could occur. Does this mean that in a cathodic reaction the electron jumping to the oxidized species (e.g. Fe^{3+}) is slow?*

The speed or slowness of an electrode reaction depends, as in the case of any 'purely chemical' reaction, on its activation energy which the reactant must overcome to become the reaction product. This dependence is expressed in the generally valid Arrhenius relationship, which, for example, for an anodic reaction is written as

$$k_{ox} = k_{ox}^0 \exp\left(-\frac{\Delta \tilde{H}_a}{RT}\right) \tag{1}$$

where k_{ox}^0 is a potential independent constant and $\Delta \tilde{H}_a$ the activation energy of the reaction. The theory of a simple electrode (or electron transfer) reaction is

---

* Many years ago this was a puzzle to Professor J. Heyrovský who was not satisfied with the basic issues of electrochemical kinetics. 'How can a slow jumping of electrons be possible when the electrons move with the velocity of light!' was his objection. However, the slowness of an electrode reaction is not caused by the velocity of electron jumping but by other processes.

based on the assumption that genuine electron transfer is indeed very fast but it is sufficiently probable only in the case that the initial as well as the final form of the system has the same configuration of atomic nuclei (the position of the electron shells can be changed). This is the so-called *adiabatic approximation* in reaction rate theory. In the present case (equation (1), page 77) the ion of iron together with its hydration shell has to be considered as the reacting system. The water envelope vibrates to and from the ion (in fact, this vibration mode is more complicated) around a definite average distance from the iron nucleus. In view of the higher charge of Fe^{3+} the average distance of its hydration sheath is lower than that of Fe^{2+}. Thus, if Fe^{3+} accepted the electron at the average distance of the hydration sheath an additional amount of energy had to be expended in order to stretch the hydration sheath to the stable position around Fe^{2+}. Under these conditions the electron transfer would be highly improbable. Therefore, the surroundings of Fe^{3+} must be reorganized so that the position of the hydration sheaths of Fe^{3+} and of the resulting Fe^{2+} are approximately equal, which is achieved in the course of vibration. In Figure 51 the energy changes of the ion surroundings are shown as functions of the distance from the iron nucleus. The intersection of the two curves indicates the coordinate of such a configuration where electron transfer is most probable. The corresponding energy value determines the activation energies of both the cathodic and the

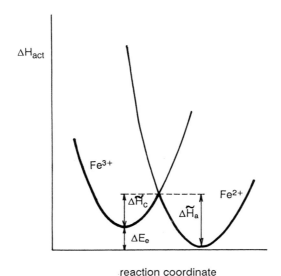

reaction coordinate

Figure 51. The dependence of the energy of the system $Fe^{3+} + e$ and Fe^{2+} on the distance of the hydration sheath from the iron nucleus. The mutual position of the curves depends on the electron energy ΔE_e (and, in this way, on electrode potential). The differences between the energy of the system at the stage of electron jump (corresponding to the intersection of both curves) and of the average energy of the two systems indicate the activation energies of the cathodic process $\Delta \tilde{H}_c$ and of the anodic process $\Delta \tilde{H}_a$

anodic reaction. If the energy of the electron is changed (this is achieved by changing the electrode potential) the two activation energies, $\Delta \tilde{H}_a$ and $\Delta \tilde{H}_c$, vary as well.

This change is expressed by simple relationships:

$$\Delta \tilde{H}_c = \Delta H_c^0 + \alpha F E \tag{2a}$$

$$\Delta \tilde{H}_a = \Delta H_a^0 - (1 - \alpha) F E \tag{2b}$$

where ΔH_c^0 and ΔH_a^0 are potential-independent constants and α is the charge transfer coefficient (in the case of a simple electron transfer $\alpha \approx \frac{1}{2}$).*

These two equations together with equations (2) and (3) on page 77 represent the theoretical basis of elementary electrochemical kinetics.

In equilibrium the current density j is equal to zero,† which does not mean that the electrode process would stop but the simultaneously proceeding cathodic and anodic reaction merely cancel each other. The current density corresponding to these fully compensating processes is called the *exchange current density*. After some tedious but elementary calculations using equations (2) and (3) from page 77 and equations (1), (2a) and (2b) the equation of the exchange current density

$$j^0 = F k^0 c_{ox}^{1-\alpha} c_{red}^{\alpha} \tag{3}$$

is obtained. The standard electrode reaction rate constant k^0 is a function of ΔH_{ox}^0, ΔH_{red}^0, α and $E_{ox,red}^0$ and is a characteristic quantity for a particular electrode reaction.

The existence of exchange current was directly proven by Pleskov and Miller who used radioactively labelled zinc dissolved in mercury which was in contact with a solution of non-labelled zinc sulphate (representing an isolated electrode). Because of the exchange process the labelled zinc ions appeared in the solution while their penetration from the amalgam followed equation (3).

An important concept of electrochemical kinetics is the *overpotential* (overvoltage) already mentioned on page 93. It is defined as the difference between the actual potential of the electrode E and of the equilibrium potential $E_{ox,red}$ given by the Nernst equation:

$$\eta = E - E_{ox,red} \tag{4}$$

According to the discussion on page 77 the measured current density j can be considered as a sum of separate anodic and cathodic current densities (by

---

* A more exact relationship for these activation energies is the Marcus equation, for example

$$\Delta \tilde{H}_c = \frac{(H_{reorg} + F E)^2}{4 \Delta H_{reorg}}$$

where ΔH_{reorg} is the *reorganization energy*. This equation gives the above relationship for $\Delta H_{reorg} \gg F E$ and for $\alpha = \frac{1}{2}$.

† Using equations (2) and (3) from page 77 and equations (1), (2a) and (2b), the Nernst equation (page 84) can easily be deduced for $j = 0$.

definition the cathodic current is negative):

$$j = j_a + j_c$$

Both of these partial current densities are governed by equations, the deduction of which is indeed complicated but can be carried out with some effort using equations (2) and (3) on page 77 and the above equations, (1) to (4):

$$j_a = j^0 \exp \frac{(1 - \alpha)F}{RT}, \qquad j_c = j^0 \exp\left(-\frac{\alpha F \eta}{RT}\right) \tag{5}$$

Obviously, when increasing η, j_a becomes much larger than j_c and vice versa for the case of large negative η. Thus, one of these terms can be neglected in comparison with the other for reasonably large η ($|\eta| > 100$ mV is quite sufficient). The relationships $j \approx j_a$ and $j \approx j_c$ are called the Tafel equations for irreversible electrode reactions. The polarization curve of hydrogen evolution (see Figure 50a) is adequately described by the Tafel equation for a cathodic reaction. In the Tafel equation only one of the concentrations of the electroactive substance, oxidized or reduced, appears, which can be shown by inserting for η from equation (4). The result for the cathodic process reads

$$j = -Fk^0 c_{ox} \exp\left[-\frac{\alpha(E - E^0_{ox,red})}{RT}\right] \tag{6}$$

The problem of decreased or increased concentrations of electroactive substances in the neighbourhood of the electrode was already discussed on pages 80 and 100. It can be quantitatively treated in a straightforward way when the transport of the electroactive species to and from the electrode is controlled by convective diffusion. Reproducible results are obtained particularly when using a rotating disc electrode (a metallic disc is inserted into the base of a cylindrical Teflon tube and rotates around the axis of the cylinder).

Consider, for example, reduction of 10^{-3} M Fe^{3+} in 1 M H_2SO_4 on a rotating disc electrode (Figure 52). Owing to the reduction process the original

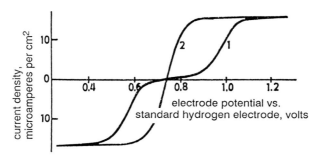

Figure 52. Voltammograms of 2.7×10^{-3} M Fe^{3+} and 2.4×10^{-3} M Fe^{2+} in 0.5 M $HClO_4$. Rotating gold electrode, rotation speed 2300 min^{-1}, chloride concentrations: 0 (curve 1) and 10^{-5} (curve 2) mol dm^{-3}. (According to J. Weber, Z. Samec and V. Mareček)

concentration of Fe^{3+}, c_{ox}^0, decreases to a value at the surface of the electrode, c_{ox}^*. The flux of the substance to the electrode is given by modifying the equation on page 63:

$$J = \frac{j}{F} = -D_{ox} \frac{c_{ox}^0 - c_{ox}^*}{\delta_{ox}} \tag{7}$$

When the overpotential is sufficiently increased the value of c_{ox}^* approaches zero and the limiting current density is given by the relationship

$$j_{lim} = -FD_{ox} \frac{c_{ox}^0}{\delta_{ox}} \tag{8}$$

Now substitute c_{ox}^* for c_{ox} in equation (6) and combine the three equations (6), (7) and (8). The result is the equation of the voltammogram (wave) due to an irreversible electrode reaction :

$$E = E_{1/2} + \frac{RT}{\alpha F} \ln \left(\frac{j_{lim} - j}{j} \right)$$

where $E_{1/2} = E_{ox,red}^0 + (RT/\alpha F)\ln(\delta_{ox}/FD_{ox})$ is the half-wave potential of the voltammogram (cf. page 101).

On adding a low concentration of sodium chloride (10^{-5} M NaCl is sufficient) the wave of Fe^{3+} reduction is shifted to more positive potentials and straightens up (see Figure 52, curve 1). As a result of adsorption of Cl^- at the electrode the standard electrode reaction rate constant considerably increases. Under these conditions not only the cathodic reaction but the anodic process as well must be taken into account. The concentrations of the reduced and of the oxidized electroactive substance at the surface of the electrode then follow the Nernst equation. Using equations (7) and (8) and analogous equations for the transport of the reduced form the reader will obtain the equation of the wave of a reversible electrode reaction. It should be noted that in this case $E_{1/2} \approx E_{ox,red}^0$.

References

1. Page xiii, Ref. 1, Chapters 3 and 4; Ref. 2, Chapter 5.
2. C. H. Bamford and R. G. Compton (eds.), *Electrode Kinetics—Principles and Methodology*, Elsevier, Amsterdam, 1986.
3. R. G. Compton (ed.), *Electrode Kinetics*, Elsevier, Amsterdam, 1988.
4. L. I. Krishtalik, *Charge Transfer Reactions in Electrochemical and Chemical Processes*, Plenum Press, New York, 1986.
5. M. A. Newton and N. Sutin, 'Electron transfer in condensed phases', *Ann. Rev. Phys. Chem.*, **35**, 437 (1984).
6. P. Sorinaga (ed.), *Electrochemical Surface Science: Molecular Phenomena at Electrode Surfaces*, American Chemical Society, Washington, D.C., 1988.

HOW TO INCREASE THE RATE OF AN ELECTRODE PROCESS

Adsorption is the accumulation of a species on the surface between two media or phases (an interface). Adsorption at the metal/gaseous phase or metal/liquid phase interface is characterized according to the type of bond between the metal and the adsorbed substance. Rather weak physical adsorption is due to van der Waals forces. Then the particles are bound to the metal surface as well as released from it (desorbed) at a high rate. In chemisorption a chemical bond is formed between the metal surface and the adsorbed particle (the adsorbate). This bond is formed and broken rather slowly. The amount adsorbed at the interface increases with the concentration of the substance in the adjacent medium as long as the interface is not fully saturated by the adsorbate. The dependence of the amount adsorbed per unit area of the interface on the concentration is termed the adsorption isotherm.

In the experiment on page 106 the electrons not only 'jump' from the metal to ferric ions but are transferred through the adsorbed chloride ions (the 'electron bridges'). In this way the electron transfer is made easier or, in other words, the electrode reaction is catalysed. The acceleration of an electrode reaction as a result of adsorption of catalysts or of reactants, reaction intermediates or reaction products is termed *electrocatalysis*. Since various metals adsorb the participants of an electrocatalytic reaction with different strengths, the catalytic effect strongly depends on the electrode material. Various kinds of electrode pretreatment (e.g. by anodic oxidation under surface oxide formation, by cathodic reduction of an oxide layer, etc., in the simplest cases) strongly change the catalytic properties of the surface.

A typical example of an electrocatalytic process is the electrolytic hydrogen evolution

$$2 H_3O^+ + 2 e = H_2 + 2 H_2O$$

which was for a number of decades the electrode reaction most studied in theoretical electrochemistry.* The hydrogen overpotential was first defined as the value of the voltage imposed on electrodes when hydrogen bubbles just started to evolve. In 1905 J. Tafel formulated the quantitative relationship between the overpotential of hydrogen ion reduction and the current density. In the subsequent systematic research it was shown that the overpotential strongly depends on the cathode material, and the actual course of the hydrogen evolution is more complicated than might follow from equation (5) on page 105 because it is a typical electrocatalytic process. Gaseous hydrogen alone reacts with various metals in the way that its molecules dissociate to atoms that are

---

* This was the period when electrode processes were considered at the end of textbooks or chapters on electrochemistry and where no quantitative aspects but rather only vague hypotheses on the course of electrode processes were presented.

subsequently bound to the metal by a chemisorption bond according to the equation

$$H_2 \rightleftharpoons 2 H_{ads}$$

that is $\qquad\qquad H_2 + \text{surface} \rightleftharpoons 2(H \ldots \text{surface})$

When changing the type of metal the equilibrium between gaseous hydrogen molecules and adsorbed hydrogen atoms is shifted to different degrees to the side of adsorbed hydrogen; in other words, different metals bind hydrogen with different strengths. Negligible hydrogen adsorption is characteristic of mercury, lead, thallium and cadmium. A medium adsorption effect is observed for the platinum group metals. Very strong adsorption is exhibited, for example, by tungsten, whose surface is almost completely covered by hydrogen atoms even if the hydrogen pressure is very low.

It is not unexpected that the adsorption of hydrogen plays a major role in electrolytic reduction of oxonium ions. In the first step in this process adsorbed hydrogen atoms are formed by the Volmer reaction

$$H_3O^+ + e = H_{ads} + H_2O \tag{1}$$

The next step is connected with the removal of adsorbed hydrogen from the electrode surface. This can proceed via the Tafel reaction

$$2 H_{ads} = H_2 \tag{2}$$

or via the Heyrovský reaction

$$H_{ads} + H_3O^+ + e = H_2 + H_2O \tag{3}$$

The strength of the adsorption bond decides (a) the slowest step that determines the overall rate of hydrogen evolution and (b) the reaction path (reaction mechanism) of hydrogen evolution.

As for mercury, the adsorption bond is weak, reaction (1) takes place slowly and requires a large additional amount of electrical energy in order to occur, i.e. the reaction needs a high overpotential. This reaction is then the slowest step in the overall electrode process, while it is not easy to distinguish between reactions (2) and (3) as a follow-up reaction, the latter being more frequently accepted as the step actually occurring. With platinum, reaction (1) takes place rather rapidly and has an equilibrium character. Because of the medium strength of the adsorption bond, recombination of hydrogen atoms (2) occurs with relative ease. This process is then the rate-determining step in hydrogen evolution. Tungsten binds hydrogen atoms very strongly so that breaking two adsorption bonds in order to form a hydrogen molecule would require excessively high energy. Therefore, reaction (3) occurs preferentially as only one bond, H—W, needs to be broken.

The following experiment will demonstrate the adsorption of hydrogen on a platinum electrode. In the electrolytic cell shown in Figure 49, 0.5 M H_2SO_4 is electrolysed with a platinum electrode using cyclic voltammetry. The resulting

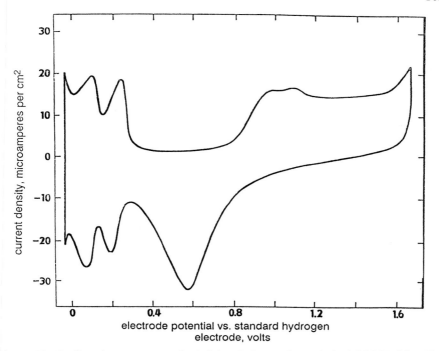

Figure 53. Cyclic voltammogram of a bright platinum electrode in 0.5 M H_2SO_4. The rate of scanning was 30 V s^{-1} and oxygen was removed from the solution by argon bubbling. (By courtesy of J. Weber)

voltammogram is shown in Figure 53. Two peaks in the potential range of 0.0–0.2 V correspond to the formation of an adsorbed layer of hydrogen atoms (lower part of the diagram) and to oxidation of this layer to hydrogen ions (upper part). The cause of the formation of two peaks lies in the different energies of hydrogen adsorption (strength of adsorption bond) at different planes of polycrystalline platinum. The peaks in the more positive potential range correspond to oxidation of platinum to a surface oxide (upper part of the diagram) and to reduction of this oxide (lower part).

Electrolytic hydrogen evolution is important for the production of very pure hydrogen. It is also supposed to be the basic step in the 'hydrogen economy'. In this, still rather hypothetical, project for utilization of nuclear energy, the heat gained from the nuclear pile is transformed into electrical energy in the conventional way. The electrical energy would not then be fed into the electrical grid but would be used for hydrogen production. Instead of electrical current, hydrogen distributed by pipelines in a similar way to transportation of natural gas would then become the main energy carrier. Incidentally, the cost of such a pipeline system is comparable to the cost of the high-voltage transmission system. The local conversion of hydrogen energy to electricity would occur in gas turbines, in fuel cells (see page 97) or by direct burning. Personally, I do not think that the 'hydrogen economy' would become a general energy system in the future while, on the other hand, pure hydrogen, the production and storage of which has achieved considerable improvement, particularly in connection with the space programme, will be rather widely used in future technologies as a powerful and clean reductant.

In the production of heavy water, D_2O, electrolytic hydrogen evolution also plays an important role. The gaseous hydrogen formed by electrolysis contains a higher proportion of light hydrogen than the original water so that the solution is enriched with respect to heavy hydrogen. This enrichment is advantageous as a preliminary step in the overall separation of heavy water.

References

1. Page xiii, Ref. 6, Volume 8.
2. A. N. Frumkin, 'Hydrogen overvoltage and adsorption phenomena, Part I', in *Advances in Electrochemistry and Electrochemical Engineering* (eds. P. Delahay and C. W. Tobias), Vol. 1, Wiley-Interscience, New York, 1961, p. 65; 'Part II', Vol. 3, 1963, p. 267.
3. L. I. Krishtalik, 'Hydrogen overvoltage and adsorption phenomena, Part III', in *Advances in Electrochemistry and Electrochemical Engineering* (eds. P. Delahay and C. W. Tobias), Vol. 7, Wiley-Interscience, New York, 1970.

ELECTROCHEMICAL DREAM: COLD NUCLEAR FUSION

Some metals (Pd, Fe, La, Ce, Ti, Zr, Th, V, Nb, Ta) when used as a cathode not only produce molecular hydrogen but also dissolve hydrogen. Hydrogen brittling of iron is one of the results. The ability to dissolve hydrogen is strongest with palladium where atomic ratios (Pd:H) as high as 1:1 can be reached. In March 1989, M. Fleischmann and S. Pons claimed that during electrolysis of heavy water at a palladium electrode the nuclear fusion of two deuterium atoms took place accompanied by conspicuous evolution of heat. They also described a certain, even if not very strong, neutron and γ-ray emission during the electrolysis. Their announcement evoked a wave of enthusiasm as well as of criticism. Obviously, the possibility of having a new energy source using very simple means could be compared with purposeful use of fire in the Stone Age. Unfortunately, the present opinion on 'cold nuclear fusion', as it was called, is prevailingly sceptical.

MORE ELECTROCATALYSIS

In a similar way as in the hydrogen evolution reaction, oxygen reduction also consists of several reaction steps. On some electrode materials oxygen is first reduced to hydrogen peroxide, which is then reduced to water at more negative potentials:

$$O_2 + 2 H^+ + 2 e = H_2O_2$$

$$H_2O_2 + 2 H^+ + 2 e = 2 H_2O$$

Oxygen reduction occurs at a mercury electrode via this two-step process. In fact, oxygen is reduced to hydrogen peroxide here in a series of partial steps:

$$O_2 + e = O_2^-$$

$$O_2 + H^+ = HO_2$$

$$HO_2 + e = HO_2^-$$

$$HO_2^- + H^+ = H_2O_2$$

but no further complications occur in the absence of a catalyst. A somewhat different situation is encountered with a silver electrode where the presence of a silver oxide on the electrode surface has a strong impact on the reaction mechanism. Hydrogen peroxide reduction then occurs at more positive potentials than reduction of oxygen to hydrogen peroxide. As a result, oxygen is reduced directly to water.

In the absence of the oxide, the reduction of hydrogen peroxide takes place in the same way as at a mercury electrode.

When ferrous ions are present in the solution during oxygen reduction they react with the hydrogen peroxide formed at the electrode during its diffusion from the electrode:

$$Fe^{2+} + H_2O_2 + H^+ = Fe^{3+} + H_2O + \dot{O}H$$

$$Fe^{2+} + \dot{O}H = Fe^{3+} + OH^-$$

The ferric ions produced in this way are again reduced at the electrode and, in this cyclic process, catalyse oxygen reduction to water. Sometimes, in a more complicated way, but usually with a much more profound effect, catalytic electrochemical reductions of oxygen occur in the presence of metal complexes, particularly of haemoproteins* like haemoglobin or catalase or their prosthetic group hemin.

Oxygen evolution, i.e. the process $4 OH^- = O_2 + 2 H_2O + 4 e$, always takes place with a considerable overpotential and moreover at 'noble metal' electrodes, i.e. at metals that do not oxidize at a lower potential than that of oxygen

---

* Haemoproteins are enzymes containing a porphyrin complex of iron as their active site (a prosthetic group) (see page 90).

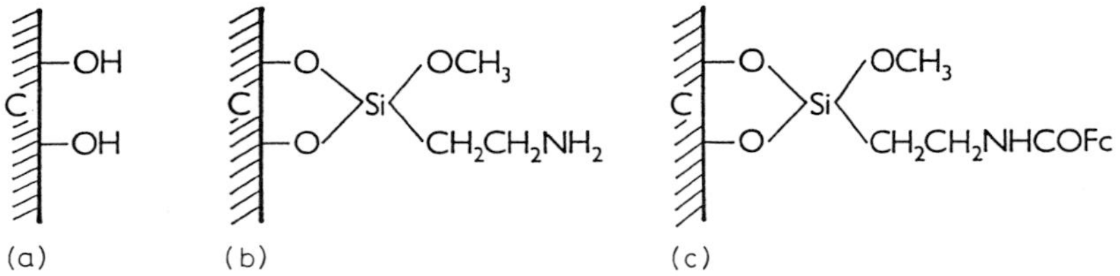

Figure 54. Chemical modification of a graphite electrode. The surface is first silanized and then a ferrocene carboxylic group containing the redox system is bound to it

evolution. The reaction products are probably adsorbed hydroxyl radicals which are subsequently oxidized to adsorbed oxygen formed by recombination of O_{ads}.

The electrocatalysis discussed so far was concerned with participation of adsorption of catalytically active substances present in solution or with a suitable electrode material. A remarkable type of catalytic electrodes are *chemically modified electrodes*. This modification is intended to enhance the selectivity of the electrode process of the substance of interest or to suppress the electrode process of the interfering species. For example, bound hydroxyl groups were found on the surface of graphite. According to Kuwana and Murray these groups are silanized (i.e. coupled with derivatives of hydrogen silicide, silane). Suitable redox groups are then linked to the resulting layer of bound siloxane groups. These redox groups then mediate the electron transfer to the electrode (see Figure 54).

Reference

1. A. Merz, 'Chemically modified electrodes', in *Electrochemistry* (ed. E. Steckhan), Vol. IV, Springer, Berlin, 1989.

ELECTROCHEMICAL TECHNOLOGIES

Electrolytic evolution of hydrogen and oxygen is an important process of preparation of pure hydrogen. However, technical hydrogen is mostly produced by the catalytic reaction of hydrocarbons (methane in the first place) with water vapour or is isolated from water gas (the product of the reaction of coke with water vapour). The most important electrochemical technologies are the electrolysis of brine (sodium chloride solution) whereby chlorine and sodium hydroxide are produced, aluminium production and electrochemical plating (particularly of automobile parts).

Brine electrolysis is based on anodic oxidation of chloride ions and on cathodic hydrogen evolution.* The two electrodes are separated by an ion-

---

* The traditional technology used a mercury cathode where the sodium ion was reduced to obtain sodium amalgam which was then decomposed by a corrosion mechanism (see page 119) to sodium hydroxide and hydrogen. This technology is no longer in use, particularly because of the toxicity of mercury.

titanium
substrate

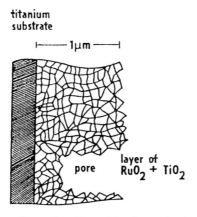

Figure 55. Beer–De Nora dimensionally stable electrode. A titanium support plate is covered with a very thin porous layer consisting of a RuO_2-TiO_2 mixture

exchanger membrane NAFION (see Table 3) which is permeable to sodium ions only. Chlorine is evolved at ruthenium dioxide anodes of the Beer–DeNora system (Figure 55). In the process various oxidation degrees of ruthenium play a role:

$$Ru^{IV} \rightleftarrows Ru^{V} + e$$

$$Ru^{V} + Cl^{-} \longrightarrow Ru^{V}Cl + e$$

$$Ru^{V}Cl + Cl^{-} \longrightarrow Ru^{IV} + Cl_2$$

Hydrogen evolution takes place on a stainless steel cathode. A scheme of this electrolytic cell is shown in Figure 56.

The Heroult–Hall process of aluminium production is based on electrolysis of a molten electrolyte consisting of pure Al_2O_3 and of cryolite Na_3AlF_6 with

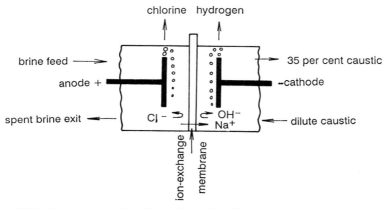

Figure 56. Scheme of the brine electrolytic cell with the ion-exchanging membrane

114

AlF$_3$, or also CaF$_2$ admixture. At a graphite anode O$_2$ and CO$_2$ are produced (the anode is thereby consumed) and liquid aluminium acts as the cathode. This process is energetically rather disadvantageous and ecologically objectionable, but no viable technological alternative has been found in spite of the fact that the patent covering it was granted in 1886.

References

1. E. Heitz and G. Kreysa, *Principles of Electrochemical Engineering*, Verlag Chemie, Weinheim, 1986.
2. M. I. Ismail (ed.), *Electrochemical Reactors: Their Science and Technology*, Elsevier, Amsterdam, 1989.
3. M. Landau, E. Yeager and D. Kortan (eds.), *Electrochemistry in Industry: New Directions*, Plenum Press, New York, 1982.
4. I. Roušar, K. Micka and A. Kimla, *Electrochemical Engineering*, Vols. 1 and 2, Elsevier, Amsterdam, 1985.

SYNTHESIS BY ELECTROLYSIS

Organic compounds are much more numerous than inorganic species, those already known exceeding four million in number. The number of electrode reactions of organic substances compared to that of inorganic processes also exists in the same proportion. On the cathodes of various materials there are reduced double bonds between carbon atoms, in particular conjugated double bonds, the nitro, halogen and carbonyl groups, pyridine and other heterocyclic nuclei, etc. At anodes, not only simple oxidations but also chlorinations and cyanations take place, and even carboxylic acids can be converted to hydrocarbons and carbon dioxide in the Kolbe reaction (in fact, first described by M. Faraday).

Electrosynthetic methods have some advantages, for example the yield of an electrochemical reaction taking place at constant potential is often large, but they are restricted to laboratory work. In industrial production they have two main drawbacks: they proceed rather slowly since they are confined to the electrode/electrolyte solution phase boundary and they consume expensive electric energy. The only large-scale electrochemical process in the field of organic chemistry is reductive dimerization of acrylonitrile to adiponitrile in the sequence of reactions

$$CH_2{=}CHCN + 2\ e = \bar{C}H_2{-}\bar{C}HCN$$
$$\text{acrylonitrile}$$

$$\bar{C}H_2{-}\bar{C}HCN + CH_2{=}CHCN = NC\bar{C}HCH_2CH_2\bar{C}HCN$$

$$NC\bar{C}HCH_2CH_2\bar{C}HCN + 2\ H_2O = NC(CH_2)_4CN + 2\ OH^-$$
$$\text{adiponitrile}$$

(In a dimerization two molecules of the reactant, a monomer, are converted into one molecule, a dimer.) Adiponitrile is the basic material in the synthesis of Nylon 606.

References

1. M. M. Baizer and H. Lund (eds.), *Organic Electrochemistry: An Introduction and a Guide*, 2nd ed., Marcel Dekker, New York, 1983.
2. A. J. Fry, *Synthetic Organic Electrochemistry*, John Wiley & Sons, Chichester, 1989.
3. D. K. Kyriacou and D. A. Jannakoudakis, *Electrocatalysis for Organic Synthesis*, Wiley–Interscience, New York, 1986.
4. T. Shono, *Electroorganic Chemistry as a New Tool in Organic Synthesis*, Springer-Verlag, Berlin, 1984.
5. K. Yoshida, *Electrooxidation in Organic Chemistry, The Role of Cation Radicals as Synthetic Intermediates*, John Wiley & Sons, New York, 1984.

ELECTROLYTIC DEPOSITION AND DISSOLUTION OF METALS

Deposition of metal-forming amalgams at mercury electrodes is a relatively simple process with properties resembling those of a simple redox process.

In a more complicated manner metals are deposited on other solid electrode materials (mercury on platinum or platinum on graphite, for example) or on electrodes made of the same solid metal. These processes are governed by the laws of formation and growth of a new phase. When solid metals are deposited the whole process is called electrocrystallization.

The basic rule for solid metal deposition (as for any crystallization) requires the particles to be deposited most easily in steps on the crystal surface where one layer of the crystal lattice covers the underlying layer and, particularly, at kinks formed in these steps (see Figure 57a). For a newly deposited metal atom the energetically most favourable position is that with the greatest contact with the metallic phase already present. This is achieved at a kink in the step (called the half-crystal position, HCP).

Such a structure of metal surface was in good accord with the theory of electrocrystallization but all this was only surmised and direct proof was missing. Only the scanning tunnelling microscope (STM) made it possible to 'see' the steps and kinks directly in solution (Figure 57b).

Crystal growth is made easier by the presence of screw dislocations in the crystal lattice. Then the step on the crystal surface does not have rectangular form as the higher atomic layer is slightly sloping so that it enters the next lower layer (see Figure 58). This strange sort of step remains on the surface of a crystal during crystallization so that it rotates around the point where it disappears in the layer above. The crystallization then occurs in a characteristic spiral manner. As a result, a low-sloped pyramid of the deposit is formed.

116

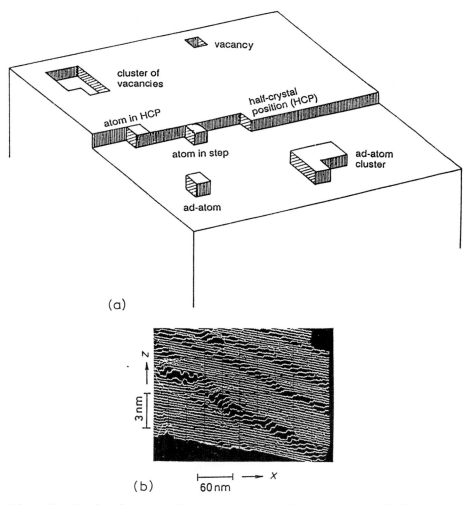

Figure 57. Metal surface on molecular scale. (a) A schematic picture with little cubes representing individual atoms. (According to E. Budevski.) (b) Morphology of a flame-treated gold electrode (crystallographic face 111) in 0.05 M H_2SO_4 + 5 mM NaCl. The image obtained with scanning tunnelling microscopy (STM) has a × 20 higher magnification in the perpendicular direction to the face than in the parallel direction. (By courtesy of D. M. Kolb)

At sites where more numerous steps and dislocations were present on the original metal surface, the metal is deposited very intensely. Coarse grains (crystallites) are formed and, sometimes, the deposit grows in shrub-shaped aggregates (dendrites, from Greek *dendron* = tree; see Figure 59). In industrial metal deposition these phenomena are rather unwelcome, particularly in electrolytic plating. Therefore, rapid local growth of the deposited metal can be

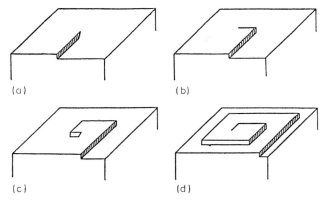

Figure 58. A scheme of the spiral growth of a crystal. (a) A screw dislocation. (b) Rows of atoms are deposited in the step so that a new step is formed. (c) The process of (b) is repeated and the third and fourth step is formed. (d) In all steps now atom rows are deposited resulting in a growth spiral. (According to R. Kaischew, E. Budevski and J. Malinovski)

inhibited by various additives that adsorb on active sites of the electrode. The resulting deposit then has a microcrystallinic, consistent structure and its surface attains a bright lustre. Sometimes the growth of a small crystal in only one direction can be achieved using additives. Then a very long and very hard crystal, a 'whisker', is formed.

Anodic oxidation of solid metals follows similar laws as electrocrystallization. It occurs most easily at vacancies, steps and dislocations in the metal surface

Figure 59. Dendritic growth during electrocrystallization of lead. (By courtesy of A. Despić)

(Figure 57a). The products are either metal ions dissolved in the electrolyte or insoluble species, particularly oxides, which remain on the surface as films. These films (mainly when they are porous) often make further oxidation possible. Sometimes, however, particularly with chromium, nickel, tantalum, titanium, aluminium and, under certain conditions, iron, they form a compact layer and bring the metal into a *passive* state. The resulting oxide film is quite often a solid electrolyte with a very low conductivity and protects the metal from further oxidation.

Oxide films formed on metal surfaces differ both in thickness and structure. Continuous films are very thin. Their thickness is between nanometres and micrometres while porous films reach millimetre thickness. Very thin films of molecular dimensions and other similar layers adsorbed on electrodes are often investigated using various non-electrochemical, physical methods. Some of these are applicable directly in the electrolysed solution (*in situ* methods) while others working at high vacuum require the electrode to be removed from the solution and dried (*ex situ* methods). The first group includes various optical methods— in the first place those analysing electromagnetic radiation reflected from the surface of the electrode (in the visible, ultraviolet and infrared wavelength range) or dispersed (Raman) radiation. Various methods based on analysis of electron or ion beams or X-rays formed by the action of electrons, ions and X-rays on the surface of the electrode belong to the second group.

An interesting application of anodic metal dissolution is electrochemical machining, discovered by Gusseff in 1928 but introduced into technology about 1960. Here, the cathode of the cutting tool is kept at a minimum possible distance from the machined material (which is an anode) so that, after switching on the electric current, a mirror image of the tool is formed in the material by anodic dissolution. The electrolyte solution flows rapidly through the narrow space between the anode and the cathode to decrease the overheating of the electrode and to remove the products of the anodic process. Both parts are kept at constant distance in order to compensate for the loss of the anode. In this way arbitrary profiles can be formed in the machined material. The second advantage, in comparison with traditional machining, is that a great variety of metallic materials can be machined since this method does not depend, for example, on the metal hardness. A scheme of electrochemical drilling is shown in Figure 60.

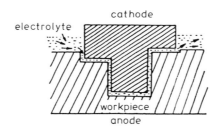

Figure 60. Electrochemical machining

Metal deposition from its coordination compounds (complexes) requires some additional electrical energy which corresponds at least to the energy of formation of the complex of its components, the free metal ions and the coordinated particles (ligands). The participants in the electrode reaction are often the particles of the complex while sometimes the complex first has to dissociate to its components. This reaction may not be fast enough and its course then determines the rate of the process.

References

1. Page xiii, Ref. 6, Volume 7 and 8.
2. Page xiii, Ref. 2, Chapter 5.

CORROSION

As already mentioned on page xi, a negative role that electrochemistry plays in the world economy belongs to the field of corrosion. There are many mechanisms of corrosion of metallic materials, electrochemical as well as non-electrochemical. An example of a simple process of electrochemical corrosion is demonstrated by the following experiment.

In the three-electrode electrochemical cell (Figure 49) the polarized electrode is substituted with a layer of dilute zinc amalgam poured on the bottom of the cell. At the same time, a mechanical stirrer with changeable rotation speed is placed in the amalgam. After the amalgam comes into contact with 0.1 M H_2SO_4 solution poured into the cell, hydrogen bubbles immediately appear on its surface. Simultaneously zinc ions are set free in the solution. From this experiment a layman will conclude that a chemical reaction,

$$Zn + 2 H_3O^+ \longrightarrow Zn^{2+} + H_2 + 2 H_2O$$

proceeds in the cell. However, when measuring the electrode potential of the amalgam, a definite value, E_{mix}, is found. An increase of stirring accelerates hydrogen evolution as well as formation of zinc ions in the solution and the relevant potential is designated E'_{mix}. At this stage, the system will be altered by substituting pure mercury for zinc amalgam. When a polarization curve is recorded with this electrode curve 1 in Figure 61 (cf. Figure 50a) is obtained. The rotation of the stirrer exerts no influence on this dependence since the electrode reaction is irreversible and slow, as already noted on page 100. In another modification of the experiment the zinc amalgam is polarized in a neutral solution of sodium sulphate. The resulting polarization curve 2 of zinc oxidation (ionization) is shifted at an increased rotation speed to more positive potentials (curve 2'). Now the same absolute values of current are found on polarization curves 1 and 2 and 1' and 2' and the corresponding potentials are determined. Obviously, these potentials coincide with E_{mix} and E'_{mix}. Thus, the hydrogen evolution and the zinc oxidation are not caused by the above 'chemical' reaction

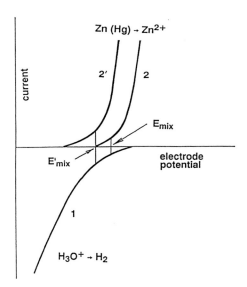

Figure 61. The origin of the mixed potential. Curve 1 represents the discharge of hydrogen ions at a mercury electrode in 0.1 M H_2SO_4. Electrochemical oxidation of the zinc amalgam in a neutral solution results in curves 2 (lower stirring rate) and 2' (higher stirring rate). E_{mix} and E'_{mix} are the corresponding mixed potentials

but are due to individual electrode reactions $Zn(Hg) \rightarrow Zn^{2+} + 2e$ and $2 H_3O^+ + 2e \rightarrow H_2 + 2 H_2O$. Evidently the zinc dissolved in mercury has no effect on hydrogen evolution and, at the same time, the change in pH does not influence the oxidation of amalgamated zinc. The process of oxidation of a metal in a medium where the rate of oxidation is compensated by reduction of hydrogen ions (reduction of oxygen will have a similar function) is called *electrochemical corrosion*.

Passive films on the surface of metals (see page 118) protect them against corrosion. This is the reason why such alloys of iron are chosen where appropriate films are readily formed, for example by the influence of nickel or chromium admixtures. Protective films can be obtained by treating the surface with phosphoric acid. In addition to slowing down the dissolution rate, these films prevent the electroactive species in the cathodic reaction (oxonium ions, oxygen) from reaching the surface by diffusion or inhibit their electrode reactions. The coating of the surface by organic materials, particularly of polymeric character, is carried out for similar reasons.

In the organic chemistry laboratory various compounds are reduced by metals or metal amalgams. These reactions are often considered as 'purely' chemical, in the same way as in the case of the corrosion of zinc amalgam. However, even here a corrosion mechanism is displayed; for example when nitrobenzene is reduced by zinc powder, its molecules are reduced by electrons from the metal and, at the same time, the zinc ions are formed by anodic oxidation of zinc.

References

1. M. Froment (ed.), *Passivity of Metals and Semiconductors*, Elsevier, Amsterdam, 1983.
2. E. McCafferty and R. J. Brodd (eds.), *Surfaces, Inhibition and Passivation*, The Electrochemical Society, Pennington, 1986.
3. E. Mattsson, *Basic Corrosion Technology for Scientists and Engineers*, Ellis Horwood, Chichester, 1989.
4. R. Štefec, *Corrosion Data from Polarisation Measurement*, Ellis Horwood, Chichester, 1990.
5. H. H. Uhlig, *Corrosion and Corrosion Control: An Introduction to Corrosion Science and Engineering*, 2nd ed., John Wiley & Sons, New York, 1971.

ELECTRODES: THEIR INTERFACIAL STRUCTURE

On page 63 the Nernst or diffusion layer is discussed which is a structure formed, in some cases, during electrolysis at the surface of the electrode. Even in extreme cases (very rapid stirring, application of ultramicroelectrodes; see page 131) the diffusion layer thickness can hardly be diminished below the order of micrometres. In the present section some insight into the much closer vicinity of the electrode will be attempted, to distances amounting from tens to tenths of nanometres. This region is called the *electric double layer*. The term double layer originates from the idea that electric charge in the electrode (already mentioned on page 77) and the opposite charge in the solution are in a certain sense ordered to form a molecular condenser ('molecular' is used here because of the molecular dimensions of the separation of the charged layers).

The properties of the electric double layer will be demonstrated using a similar electrode system as in the corrosion experiment (page 119). There are several reasons (high hydrogen overpotential, easy purification of the surface, absence of oxide films) why a mercury electrode is most suitable for double-layer investigations. In the present experiment the electrolyte solution contains 0.1 M sodium sulphite and 0.1 M NaOH. Pure argon or nitrogen bubbles through the solution while the mercury electrode is polarized by a voltage of -0.7 V as a cathode; the saturated calomel electrode is the anode. In this way a small amount of mercuric ions formed in the solution during contact with mercury in the presence of traces of atmospheric oxygen, which was not reduced by the sulphite, is reduced to metallic mercury. After some time the electric current falls to zero. The solution then contains only sulphite, hydroxide and sodium ions together with a small concentration of sulphate ions formed by oxidation of sulphite by oxygen. Under these conditions the system remains in a currentless state. When the voltage is increased to -1.0 V a large instantaneous current flows through the circuit and immediately falls to zero. After that a potential of -1.0 V is maintained with no further supply of electric charge. The initial current obviously is the charging current (see page 77). The negative potential can be increased to a value of -1.2 V with the same result—on charging from an external source the electrode acquires a definite potential value which is maintained for an arbitrary period of time. Thus, it behaves like a condenser without leakage. It is called an *ideally polarized electrode*.

A similar experiment could be carried out with a potassium chloride or potassium fluoride solution but the difficulties with removing oxygen from the solution prevent it being carried out without tedious precautions. Oxygen is reduced at the electrode (naturally, also in the situation when the external voltage source is disconnected) and discharges the molecular condenser (it decreases the negative charge of the metal by removing the electrons). When no current is supplied from the external source the negative potential of the electrode decreases. However, some of the properties of the electric double layer can be successfully investigated, even in the presence of trace concentrations of oxygen.

When, for example, 0.1 M KF solution completely freed of oxygen and other impurities is poured over a mercury electrode, the electrode acquires a potential of -0.19 V versus the standard hydrogen electrode. Before coming into contact with the electrolyte, the electrode possesses a negligibly small charge because the capacity of a metal in contact with a gas or vacuum is extremely low. In contact with the electrolyte (with no species present which would react at the electrode at -0.19 V), the electrode can accept no charge from it. The potential of -0.19 V versus the standard hydrogen electrode is the *zero charge potential*.

The zero charge potential strongly depends on the electrolyte composition which is connected with the ability of individual types of ions to become adsorbed at the electrode/electrolyte solution interface. Fluoride, potassium or sodium ions are not adsorbed at all. When the tendency towards adsorption is greater for the anion of the electrolyte than for the cation, the zero charge potential is more negative than in potassium fluoride or sodium fluoride solutions, e.g. in solutions of alkali metal chlorides, bromides or iodides. On the other hand, when the adsorption of the cation is stronger than of the anion (in a solution of tetrabutylammonium chloride, for example) the zero charge potential is more positive. The zero charge potential depends markedly on the electrode material (for silver, for example, it is about 0.5 V more positive than for mercury).

In order to charge the electrode to a more positive potential, a supply of positive charge is required and, similarly, a supply of negative charge for negative potentials. The amount of electricity necessary for charging the electrode depends on its capacity which, at potentials more negative than the potential of zero charge, is about 20 μF cm^{-2} in simple inorganic electrolyte solutions, while at more positive potentials it is higher. The electrode capacity is not constant but depends more or less on the potential of the electrode. As already noted, these values are very high because of the extremely small thickness of the molecular condenser. The simplest picture of the electric double layer is that of a plate condenser where one armature is represented by the charges in the metal and the other one being formed by the ions at a minimum distance in the solution (this distance is really very small, of the order of tenths of a nanometre). In this structure the electroneutrality rule is not preserved and a definite type of ion predominates in the electric double-layer region of the solution. As a whole, however, the electric double layer is electroneutral, the sum of the charge in the metallic and in the solution part of the double layer being equal to zero.

This picture of the electric double layer (introduced by H. L. F. von Helmholtz in 1879) is, however, valid for a rather concentrated electrolyte solution. The more dilute the electrolyte the larger the distances will be where the ions forming the solution part of the electric double layer can escape by thermal motion. A structure that is similar to the ionic atmosphere around individual ions (cf. page 35) then predominates. This is called the diffuse electric layer. Figure 62 shows the structure of the electric double layer at the metallic electrode/electrolyte solution interface. The thickness of the diffuse electric layer strongly depends on the electrolyte concentration. In electrolytes consisting of a

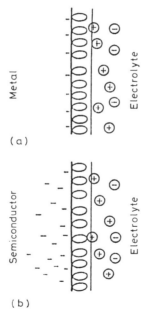

Figure 62. Electric double-layer structure. (a) A metal/electrolyte solution interface. The dashed line corresponds to the plane of ion centres at minimum distance from the electrode surface. No adsorption of the ions is assumed since otherwise the structure would be more complicated. The ovals are oriented solvent molecules (this orientation is the cause of much lower permittivity of this region than of the bulk). The overlapping of the electron cloud from the metal into the solution is not shown. (b) A semiconductor/electrolyte solution interface. Diffuse electric layers are formed in the solution as well as in the semiconductor

monovalent cation and a monovalent anion it has a thickness of an order of tenths of a nanometre at 0.1 molar concentrations, whereas the thickness becomes as high as tens of a nanometre at millimolar concentrations (cf. the very similar phenomenon, the ionic atmosphere, page 35).

The very strong electric field between layers of opposite charge in solution and at the surface region of the metal affects polar solvent molecules which become oriented along the field. This is why the permittivity of the solvent is much lower at the surface of the electrode than in the bulk. Moreover, the electron clouds from the metal may reach this part of the double layer (*compact electric layer*) forming a structure termed *jellium*.

Ideally polarized electrodes can be attained in few electrode systems, particularly with regard to the electrode metal. On the other hand, electrodes on which an equilibrium potential is established have an electric double layer of the same character. This is an important fact, for example, for semiconductor electrodes which are corroding systems in most media. A change in the electric double-layer properties is then achieved through a change in the composition of the oxidation–reduction system present in the solution. In contrast to metallic electrodes where the charge of the metallic part of the electric double layer is homogeneously dispersed over the metal surface and the concentration of charge carriers is high, semiconductor electrodes have a diffuse electric layer even in the inside of the semiconductor, because of the limited value of its permittivity and of low charge-carrier concentration. In the terminology of solid-state physics, the diffuse electric layer is termed the *space-charge region* (Figure 62).

The electric double layer appears even at the electrolyte solution/insulator interface if the electric charges are fixed, e.g. by transfer of ions of a definite sort from the surface of an ionic crystal into the solution, by ion adsorption, by the presence of ionizable groups in polymeric materials, etc. The formation of the electric double layer has a great impact on the transport processes with participation of charged colloidal particles and of porous systems with fixed charges on the walls of the pores (cf. page 140).

Let us consider electric current flow through a porous stopper with negative charges on pure walls (Figure 63). The charges in the diffuse electric layer are set in motion and, simultaneously, the solvent is dragged with the ions so that a kind of osmosis, an *electroosmotic flow*, ensues. The more dilute the electrolyte solution the thicker is the diffuse double-layer region, so that the electroosmotic flow increases with decreasing electrolyte concentration. This flow can be stopped by a pressure difference, the *electroosmotic pressure*, between the two sides of the membrane (Figure 63).

The inverse phenomenon to electroosmotic flow appears when the electrolyte solution is pressed through the membrane in the absence of an electric current. The charge in the electrical diffuse double layers in the pores of the membrane cannot of course be dragged with the liquid, so that an electric field is formed in the pores in a direction parallel to their axis. The final result is the electric potential difference, the *streaming potential*, between the two sides of the stopper.

According to their charge, colloidal particles in an electrolyte solution move in the direction of the electric field imposed on the system or against it. This phenomenon is termed *electrophoresis*. Their velocity increases when the concentration of the indifferent electrolyte decreases. This effect shows that electrophoresis is connected with the presence of the diffuse double layer at the surface of the particles. Under the influence of the electric field the space charges in the diffuse electric layer are set in motion in the opposite direction to the movement of the particle itself. If the charge is spread to large distances from the surface, which is the case in dilute solutions, the particle exerts only a small

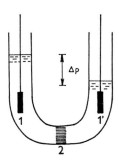

Figure 63. Electroosmotic pressure. Hydrostatic pressure difference Δp compensates the osmotic pressure difference between the compartments 1 and 1′ and prevents the solvent from flowing through the membrane 2

126

hydrodynamic resistance to the movement of the charge in the diffuse layer. On the other hand, at high electrolyte concentrations, the electric double layer is reminiscent of a plate condenser with the solution charge fixed on the plane facing the surface of the solid phase. Under these conditions it is impossible to induce a movement of the solution charge from the interface, and electrophoresis ceases. An inverse process to electrophoresis is observed when charged colloidal particles dispersed in an electrolyte solution sink to the bottom. An electrical potential difference called the *sedimentation potential* is then measured between the upper and lower layers of the solution. During sinking the diffuse electric layer lies behind the colloidal particle, resulting in an electric potential difference.

The main application of electrophoresis is the separation of proteins. In an electric field various types of these macromolecules move with different velocities because of their size and charge, so that their distribution in the electrophoretic field can be analysed optically. In preparative electrophoresis, proteins are accumulated in different compartments of the cell.

All these four phenomena, electroosmosis, streaming potential, electrophoresis and sedimentation potential, are generically called *electrokinetic phenomena*.

The electric double layer appears at the surface and, in some cases, also inside the membranes (page 140).

References

1. Page xiii, Ref. 6, Volume 1.
2. Page xiii, Ref. 2, Chapter 4.
3. A. T. Andrews, *Electrophoresis, Theory, Techniques and Biochemical and Clinical Applications*, Oxford University Press, Oxford, 1982.
4. P. Delahay, *Double Layer and Electrode Kinetics*, John Wiley & Sons, New York, 1965.
5. G. A. Marynov and R. R. Salem, *Electrical Double Layer at a Metal–Dilute Electrolyte Solution Interface*, Vol. 33, Lecture Notes in Chemistry, Springer-Verlag, Berlin, 1983.
6. S. R. Morrison, *Electrochemistry of the Semiconductor and Oxidized Metal Electrodes*, Plenum Press, New York, 1980.
7. R. D. Void and M. J. Void, *Colloid and Interface Chemistry*, Addison-Wesley Publishing Company, Reading, Mass., 1983.

ANALYSIS BY ELECTROLYSIS

Electrochemical methods of analysis of solutions have experienced periods of boom and decline. At present quite a few analytical procedures based on various sorts of electrochemical indication have survived. They may be classified by means of several criteria. When a reagent is added to the analysed solution (the sample) and an electrochemical signal (electrode potential, electric current, conductance, etc.) indicates the completion of its reaction with the substance to be determined (a determinand), this analytical approach is called a *titration method* (e.g. a conductometric titration, see page 39). Usually, the volume of reagent solution consumed in the reaction is then measured. Another group of methods is based on complete separation of the determinand by electrolysis of the analysed solution. When the amount (usually of a metal) deposited at an electrode is measured, usually by weighing, the method is termed *electrogravimetry*. *Coulometry* is based on determination of the total amount of electricity consumed for complete electrochemical transformation (reduction with or without deposition, oxidation) of the determinand. Finally, there is a group of methods that interfere with the sample only negligibly or not at all. The determinand concentration is then assessed either on the basis of a single signal or of the dependence between electrical quantities and, in some methods, of the dependence of an electric quantity on time, recorded during the electrochemical measurement. *Potentiometry* works in a currentless state making use of an equilibrium potential or of a mixed potential. Direct potentiometry is at present restricted to membrane systems termed ion-selective electrodes (see page 142). A number of electroanalytical methods are based on electrolysis with polarized *microelectrodes*. Those electrochemical methods that are of practical value at present will now be considered in greater detail.

The coulometric method requires only one reaction to occur at the electrode. For example, metallic silver is deposited on a silver electrode by electrochemical reduction of the silver ions present in the solution. According to the Faraday law (page 67), for $z_A = 0$ the electric current is directly proportional to the deposition rate of silver (i.e. to the flux of silver ions at the surface of the electrode) and, consequently, the charge consumed for silver ion reduction during a given time is proportional to the amount of silver deposited. If this amount is expressed in moles, the proportionality factor is the Faraday constant. When all the silver present in the solution has been deposited this amount can be determined according to the charge consumed for the deposition using a suitable current integrator. The coulometry at constant electrode potential is one of the most exact analytical methods.

The basic characteristics of *voltammetry* have already been pointed out on page 100. In analytical voltammetry, the dropping mercury electrode (Figure 64), the hanging mercury drop electrode and various types of rotating solid metal electrodes, particularly the rotating disc electrode, are used as indicator electrodes.

Voltammetry with the dropping mercury electrode was called *polarography* by J. Heyrovský who discovered this method in 1922. The continuous renewal of the mercury surface prevents the surfactants present in the electrolyte solution from adsorbing at the electrode, thus rendering the voltammograms reproducible. The high overpotential of hydrogen evolution at mercury extends the potential 'window', where waves of electroactive substances can be found, to quite negative potentials. In analytical polarography, generally average currents (the amount of electricity consumed during the life of a drop divided by the drop time) are recorded as a function of the imposed voltage. The average limiting

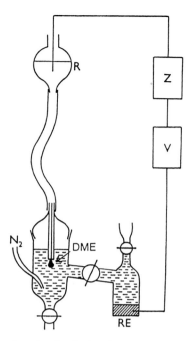

Figure 64. The scheme of a simple device for polarography with dropping mercury electrode (DME). From a glass capillary the dropping electrode drops under the pressure of mercury from the mercury reservoir R connected by means of a plastic tube to the capillary. The voltage between the dropping electrode and the reference electrode (RE) is supplied by the source V and the electric current is recorded by the recorder Z. The electrolyte solution under investigation is separated from the solution of the reference electrode by means of a wide-bore stopcock

diffusion current at the dropping mercury electrode is governed by the Ilkovič equation,

$$\text{average limiting current} = 0.63 \times 10^{-3}zFm^{2/3}t_1^{1/6}D^{1/2}c_0$$

where z is the charge number of the electrode reaction, F the Faraday constant, m the flowrate of mercury (in grams per second), t_1 the drop time (usually several seconds), D the diffusion coefficient (in square centimetres per second) and c_0 the concentration of the electroactive substance. A stationary version of the dropping mercury electrode is the hanging drop electrode where a mercury drop is squeezed out, using a micrometer screw, from a glass capillary attached to a mercury-filled glass syringe.

Polarography is a very sensitive method but the determination of substances present in micromolar and lower concentration is interfered with by the charging current, which is rather large because of the continuous increase in the surface of the dropping electrode. These concentrations (of metal ions, for example) are often of interest because they endanger the human environment. In order to increase the sensitivity of polarographic methods, various techniques have been proposed, the most successful being *pulse polarography*, which removes the interference of the charging current by measuring the current of the drop at the end of the drop life (Figure 65).

Among solid metal microelectrodes the *rotating disc electrode* introduced by A. N. Frumkin and V. G. Levich is a favourite tool in electrochemical investigation. A significant property of this electrode is the constant current density at any point of the electrode. The limiting current at the rotating disc electrode is given by the Levich equation,

$$\text{limiting current} = 0.62 \times 10^{-3} z F A D^{2/3} v^{-1/6} \omega^{1/2} c_0$$

where A is the area of the electrode (in square centimetres), D the diffusion coefficient of the electroactive substance, v the kinematic viscosity (the product of the viscosity coefficient and the density of the solution) and ω the angular velocity of the electrode.

Enormous sensitivity is achieved by *anodic stripping analysis* where a trace metal present in the sample is deposited for a definite, rather long, time on a rotating disc electrode or hanging mercury drop electrode. Then the potential of the electrode is switched to a value where the deposited metal is anodically reoxidized. The amount of electricity consumed in the reoxidation indicates with great precision the concentration of the determinand in the solution.

In some electroanalytical determinations the current is measured at constant potential. Among these *amperometric methods* the most important is the determination of oxygen in the blood and tissues. Early attempts to measure the oxygen content with a bare platinum electrode impaled into the organ were completely unsuccessful because the components of the biological fluid, particularly of a high-molecular-weight nature, were adsorbed on the electrode and inhibited oxygen reduction. These obstacles have been removed in the Clark membrane electrode (Figure 66). Here the platinum indicator microelectrode together with the reference electrode placed in a suitable electrolyte solution is

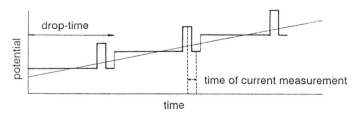

Figure 65. Voltage pulsing in differential pulse polarography. The basic slowly increasing voltage is connected to the electrolytic cell containing a dropping mercury electrode. Before the drop is torn off a small voltage pulse is imposed on the electrode. The current is integrated during the second half of this pulse and also during the identical time interval just after this pulse. The recorder indicates the dependence of this integrated current on potential

130

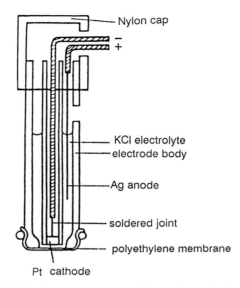

- Nylon cap
- KCl electrolyte
- electrode body
- Ag anode
- soldered joint
- polyethylene membrane

Pt cathode

Figure 66. Clark's oxygen sensor. (According to J. G. Schindler)

separated from the sample by a hydrophobic porous Teflon film. Oxygen dissolved in the sample diffuses to the electrode compartment through the gas-filled pores of the film and is reduced at the electrode.

Most of the methods discussed are suitable for analysis as well as for research on mechanisms and rates of electrode reactions. The accuracy of the coulometric method and the sensitivity of pulse polarography and of anodic stripping analysis (with a detection limit as low as 10^{-9} mol dm^{-3}) classifies them among prominent methods of analytical chemistry.

References

1. R. A. Adams, *Electrochemistry at Solid Electrodes*, Marcel Dekker, New York, 1969.
2. A. Bond, *Modern Polarographic Methods in Analytical Chemistry*, Marcel Dekker, New York, 1980.
3. G. Dryhurst, K. M. Kadish, F. Scheller and R. Renneberg, *Biological Electrochemistry*, Academic Press, Orlando, 1982.
4. Z. Galus, *Fundamentals of Electrochemical Analysis*, Ellis Horwood, Chichester, 1976.
5. J. Heyrovský and J. Kůta, *Principles of Polarography*, Academic Press, New York, 1966.
6. P. T. Kissinger and W. R. Heinemann (eds.), *Laboratory Techniques in Electroanalytical Chemistry*, Marcel Dekker, New York, 1984.
7. J. A. Plambeck, *Electroanalytical Chemistry, Basic Principles and Applications*, John Wiley & Sons, New York, 1982.
8. E. P. Serjeant, *Potentiometry and Potentiometric Titrations*, John Wiley & Sons, New York, 1984.

ULTRAMICROELECTRODES

For voltammetry, particularly in low-conductivity systems, a novel tool has recently been discovered. The ultramicroelectrode (Figure 67) is usually a tiny metallic disc (with a diameter of at most several micrometres) sealed in a glass tube (a sphere or a very thin short wire can also be used). Transport to such an electrode is not by linear diffusion but it has the basic features of *spherical diffusion*. This kind of diffusion occurs towards a spherical surface and, in contrast to linear diffusion, approaches a steady state with constant diffusion layer thickness. Similar properties are also characteristic for surfaces of other than spherical shapes, including the mentioned disc ultramicroelectrode. Voltammograms at this electrode have no peaks but are similar to those obtained with a rotating disc electrode (see page 105). The second, more important, advantage is the negligible ohmic drop of potential that makes it possible to work, for example, in solvents of low permittivity and/or at very low temperatures.

Because of its small dimension the ultramicroelectrode has fine perspectives for the determination of redox species *in vivo*. In this way it completes the application field of ion-selective microelectrodes (see page 147) which are used for analysis of ions without redox properties like those of alkaline and alkaline-earth metals. A typical example of application is the determination of neurotransmitter concentrations in nerve tissues (see page 179). The substance most

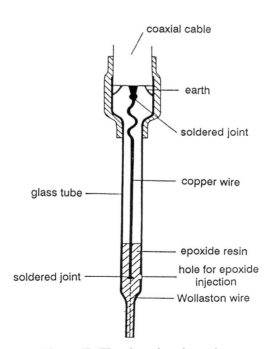

Figure 67. The ultramicroelectrode

studied so far is dopamine which is oxidized at the electrode according to the equation

$$+ 2\,H^+ + 2\,e^-$$

References

1. M. Fleischmann, S. Pons, D. R. Rolinson and P. P. Schmidt, *Ultramicroelectrodes*, Datatech Systems, Marganton, 1987.
2. J. B. Justice (ed.), *Voltammetry in the Neurosciences*, Humana Press, Clifton, 1987.

LIGHT AND ELECTRODES

As will be shown in the next chapter in more detail, the energy source for most processes on the Earth's surface is the light arriving from the Sun. The intensity of the incident light is on the average 0.14 Wcm^{-2}. The recent large-scale search for new energy sources, which would be economically acceptable and ecologically harmless, is particularly directed to the utilization of solar energy in a way that is different from that used by nature.

The incident light can be transformed to heat (with much loss of useful energy) which is then either used directly or accumulated by one of several methods. On the other hand, light may also be converted directly into electrical energy, which may also be achieved through several physical and electrochemical methods. In *photoelectrochemical energy conversion* a suitable galvanic cell is brought to an excited state by the effect of irradiation. This excitation can occur in two ways:

1. Light excites one or both electrodes of the cell. This phenomenon is termed the *electrochemical photovoltaic effect* (the main subject of this section).
2. Light affects a solution component that either directly or, more frequently, after a suitable chemical reaction changes the potential of the adjacent electrode (the *photogalvanic effect*).

Generally, in the photovoltaic effect an electric potential is formed by the irradiation of semiconductor materials. In solid-state physics these phenomena are mainly investigated with photovoltaic cells either with a *pn* junction or with a semiconductor/metal junction (Schottky-type photovoltaic cells). When the energy of the photons of the light illuminating a semiconductor plate is larger than the energy gap of the semiconductor, the electron-hole pairs form under the surface of the plate. This process cannot be utilized in an isolated semiconductor because the pair would immediately recombine, release thermal energy and the light energy would be dissipated. Photovoltaic effects that can really be observed are distinguished according to the nature of a foreign phase which is connected to the semiconductor under illumination. In electrochemical photovoltaic effects this is an electrolyte solution containing a redox system. When it is in contact with the semiconductor an electrode reaction occurs in order to equalize the Fermi levels of the electrons in the electrode and in the solution. The charging of the electrode connected with the establishment of the electrode potential results in formation of a space-charge region (see page 124). The induction of the photovoltaic effect requires a redox potential of the electrolyte such that the space-charge region is formed by the minority carriers, i.e. the holes in an n-type semiconductor and the electrons in a p-type semiconductor. Then the electric field established under the surface of the n-type semiconductor forces the electrons to migrate into the bulk of the semiconductor while the holes are transported to the surface. In this way charge separation occurs after the electron-hole pairs have been formed on illumination, which cannot be achieved in an isolated semiconductor (Figure 68). The holes thus generated charge the electrode to a more positive potential. The difference between the original equilibrium potential and this new value is called the *photopotential*. This deviation of the electrode potential from equilibrium causes the electrode reactions based on hole transfer to the reductant species in the solution to occur:

$$\text{red} + \text{h}^+ \longrightarrow \text{ox}$$

This can happen only when the process takes place in a galvanic cell; in addition to the semiconductor electrode, another, metallic, electrode connected with the semiconductor electrode through an electric energy consumer is present

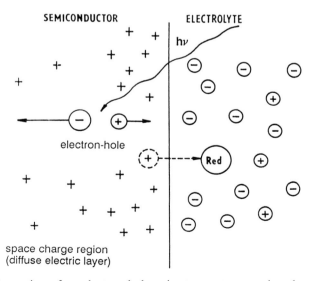

Figure 68. Formation of an electron-hole pair at an n-type semiconductor/electrolyte interface by incident light. The electric field in the space-charge region (electric diffuse layer) drives the electrons into the bulk of the semiconductor and the holes to the surface

in the solution. The electrons originally transported into the bulk of the semiconductor electrode cross the junction to the metallic electrode where an electrode reaction occurs in the opposite direction:

$$ox + e^- \longrightarrow red$$

In this way an electric current has been generated and the light energy is transformed into electric energy. Electrochemical systems containing a semiconductor electrode and a metallic electrode in a solution of a single oxidation-reduction system are termed electrochemical photovoltaic cells. They supply electric energy and their composition during service does not change since an identical electrode reaction takes place at each electrode—however, only in the opposite direction (therefore they are sometimes called regenerative photoelectrochemical cells).

The main disadvantage of these devices is the insufficient stability of the semiconductor electrode resulting from photocorrosion. The holes generated at any n-type semiconductor electrode can attack the electrode material instead of reacting with the reductant in the solution. For example, the n-CdS electrode is oxidized according to the equation

$$CdS + 2 h^+ \longrightarrow Cd^{2+} + S$$

However, this process can be suppressed by a proper choice of the redox properties of the electrolyte. The systems S^{2-}/S_n^{2-}, Se^{2-}/Se_2^{2-} and Te^{2-}/Te_2^{2-} exhibit a performance suitable for this purpose. Materials used for n-type semiconductor electrodes are sulphides, selenides and tellurides of cadmium and GaAs, InP and GaP. The efficiency of light energy conversion extends from 1 to 2% with n-CdS in a S^{2-}/S_n^{2-} electrolyte, to 10% with n-CdTe in Te^{2-}/Te_2^{2-} and from 9 to 10% with n-GaAs in Se^{2-}/Se_2^{2-}.

A photoelectrolytic cell differs from the electrochemical photovoltaic cell in that a different electrode reaction takes place at each electrode, resulting in suitable electrolysis products, in contrast to the preceding case aimed at electric energy generation.

References

1. H. O. Finklea (ed.), *Semiconductor Electrodes*, Elsevier, Amsterdam, 1988.
2. R. J. Gale (ed.), *Spectroelectrochemistry, Theory and Practice*, Plenum Press, New York, 1988.
3. H. Gerischer and J. J. Katz (eds.), *Light-Induced Charge Separation in Biology and Chemistry*, Verlag Chemie, Weinheim, 1979.
4. Yu. V. Pleskov, *Solar Energy Conversion: A Photoelectrochemical Approach*, Springer-Verlag, Berlin, 1990.

Chapter 3

Membranes

VERY THIN AND VERY THICK MEMBRANES

While the first two chapters of this book were concerned with comparatively simple structures, such as ions in solution and electrodes in electrolyte solution, the present chapter deals with electric properties of much more complicated systems that are found in living systems or simulate the properties of such systems.

The functional unit of living systems is the cell. The existence of living cells was first detected by the English naturalist R. Hooke who in 1665 drew a sketch of cork network resembling a honeycomb. The Dutch lens grinder, A. Leeuwenhoek (1628–1723), to whom the invention of the microscope is often ascribed, discovered the cells of erythrocytes, bacteria, infusoria and spermatozoa. In 1848 du Bois-Reymond suggested that the surface of a living cell has properties similar to an electrode of a galvanic cell and at the turn of the century W. Ostwald, W. Nernst and J. Bernstein predicted that living cells are enclosed by a semipermeable membrane with characteristic electric properties.

Figure 69 shows a typical animal cell. Its surface is formed by *cell* or *plasma membrane* separating the *cytoplasm* from the outer medium. The inside of the cell contains a number of corpuscles called organelles. Some of these structures are enclosed by simple membranes, others by double membranes. They display highly diverse functions including, for example, the transfer of genetic information stored as deoxyribonucleic acid (DNA) (see page 53) in the nucleus of the cell of eucaryotic organisms. The ribonucleic acid (RNA) where the information is rewritten is transported through the pores in the double nuclear membrane into the cytoplasm. Another organelle, the endoplasmic reticulum, with an intricate membrane system is one of the sites where the information involved in RNA is translated into the amino acid sequence of a protein. The mitochondria are a kind of 'floating' electric power station (for details see page 184). At their inner membranes various metabolites, such as fatty and other organic acids, are oxidized and the energy gained in the process is temporarily transduced to electric energy manifested by the change of electric potential of the mitochondrial membrane. Finally, this energy is transformed back to chemical energy of

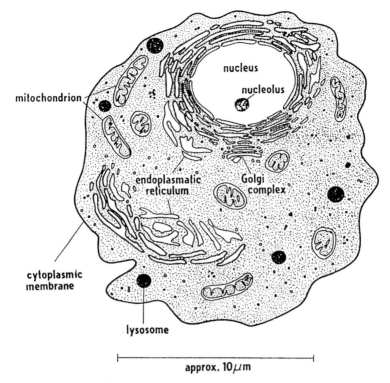

Figure 69. The scheme of an animal cell

the compound ATP (adenosine triphosphate) and exploited in metabolic functions of the organism. Miscellaneous other catalytic functions were found in the membranes of the Golgi apparatus, vacuoles, lysosomes, etc. Typical structures of green plants are chloroplasts containing thylakoids (see page 191) whose membranes mediate the transformation of radiant energy to the chemical energy of sugars.

However, what actually are the membranes? The word 'membrane' is often encountered in everyday life, in connection with telephone mouthpieces, loud-speakers or technical devices equipped with elastic plates which enable gas to flow under a definite pressure, etc. The word comes from Latin where it means parchment, i.e. leather processed in a special way so as to preserve elasticity (only on such sheets was it possible to write properly). Physicists idealized the membrane in one direction and, in acoustics, they defined it as an elastic plate of negligible thickness. The membranes that are the subject of the present chapter are, however, completely different structures.

In the second half of the nineteenth century pictures of thin tissue sections obtained with the help of the optical microscope showed that plant cells are enclosed by a distinct envelope (in the present view, a cell wall; see page 138) whereas animal cells are formed by bare protoplasm. This was the reason why W. Nernst considered the interface of two immiscible liquids as a possible model of the cell surface (there are, of course, some similarities but the relationship is rather distant; see page 142). On the other hand, in W. Ostwald's opinion the cell surface was formed by a *semi-permeable membrane*. Very often his dictum from 1891 is quoted: 'Not only electric currents in muscles and nerves but particularly even the mysterious effects occurring with electric fish can be explained by means of the properties of semi-permeable membranes.'

Ostwald also initiated research of biological membrane models which, at that time, were rather thick inorganic or organic structures (e.g. collodion film, glass bubble, etc.). This investigation found little response with the physiologists. In his monumental work *Elektrophysiologie* (Vol. I, 1895, Vol. II, 1898) the author, W. Biedermann, did not even mention the term 'membrane'.

Direct proof of the existence of a membrane on the surface of animal cells was not available before 1925 when E. Gorter and F. Grendel removed by cytolysis (in fact, plasmoptysis; see page 33) the internal content of an erythrocyte and analysed the residual matter which is now called the 'erythrocyte ghost'. For a major part it consisted of phospholipids which are derivatives of esters of fatty acids and phosphoric acid with glycerol (see page 157). When they spread the phospholipids to form a monolayer on the water surface they found that the area occupied is about twice as large as the surface of the cells. This means that the outer sheath of the cell is formed by a bimolecular layer of phospholipids. Later it was shown that all cells are covered by such a thin membrane comprising no more than two layers of molecules, electron microscopy supplying conclusive proof. As already pointed out, with plants, fungi and bacteria the situation is more complicated, because the surface of their cells, in addition to having a cell membrane immediately enclosing the cell, is reinforced by the cell wall which is a relatively thick layer formed by cellulose and other materials (Figure 70). In general, there are two basic types of membranes which differ immensely in their thickness but also have some common features. Let us therefore attempt to formulate a unifying definition of a membrane. It is a layer that separates two solutions, differs in chemical composition from both of them, forms a sharp boundary towards both these solutions and exhibits different permeability for individual components of the solution. The liquid junction of two electrolyte solutions described on page 149 is of course not a membrane. A porous body (a porous diaphragm), where one solution passes into the other one, exerts no influence on the rate of transport of individual components of the solutions and only prevents mixing by convection. We shall look at membranes through which the ions penetrate with different ease, called *electrochemical membranes*.

The physical chemistry of membranes is not restricted to modelling biological membranes but it has its own theoretical aspects as well as numerous practical applications. The branch of science comprising both theoretical knowledge and methods of investigation of artificial as well as biological membranes is sometimes called *membranology*. In this chapter various sorts of thick artificial

138

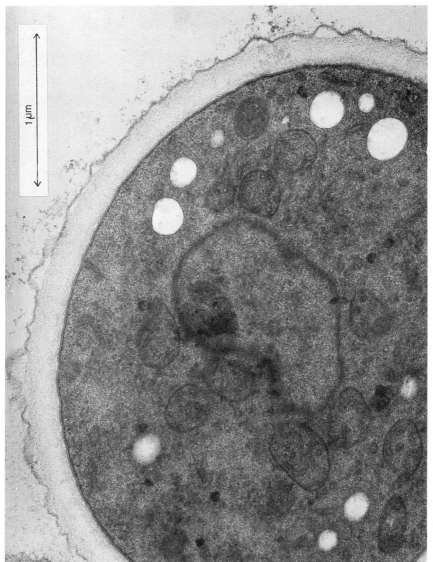

1 μm

Figure 70. Electron micrograph of a section from a yeast cell. The outer envelope is the cell wall. The inner double line is the cytoplasmatic membrane. (By courtesy of J. Ludvik)

membranes will be treated first, the characteristics of which are mostly quite far from biology, but there are some features that are identical in both groups.

References

1. Page xiii, Ref. 2, Chapter 6.
2. D. A. Butterfield, *Biological and Synthetic Membranes*, John Wiley & Sons, New York, 1989.

SYNTHETIC POLYMER MEMBRANES

This kind of membrane is also called an ion-exchange membrane. They are porous systems with pores filled with water (therefore we can also call them hydrophilic).

Consider a piece of a cation-exchanging membrane (which is commercially available) placed in a cell (Figure 71) so that it completely separates compartments 1 and 2. The cation-exchanging membrane is made of a copolymer of polyvinylchloride and sulphonated polystyrene. This material has a porous structure and the sulphonate groups, SO_3^-, are fixed on the walls of the pores (Figure 72). Compartments 1 and 2 are filled with a solution of sodium chloride with different concentrations. The ratio of the concentrations is kept constant, for example $c_2/c_1 = 10$. In both compartments are placed identical saturated calomel electrodes that have a negligible liquid junction potential at the boundary with other aqueous electrolyte solutions (page 71). The potential difference between solutions 2 and 1 measured using these reference electrodes is termed the membrane potential $\Delta\varphi_M = \varphi(2) - \varphi(1)$. At very low concentrations of sodium chloride the membrane potential is approximately $+59$ mV. When the concentration of sodium chloride increases (the ratio c_2/c_1 being kept constant) the absolute value of the potential difference decreases and, finally, at high sodium chloride concentrations it approaches about -15 mV.

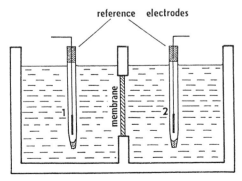

Figure 71. A cell for the study of the membrane potential

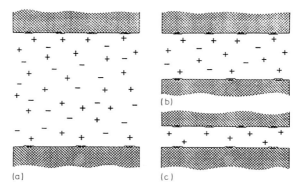

Figure 72. A scheme of a membrane pore with negative fixed charge on its surface: (a) wide pore with appreciable concentration of coions; (b) the gegenions prevail in a narrower pore; (c) the very narrow pore is completely filled with gegenions. (By courtesy of K. Sollner)

This behaviour of the membrane is closely linked to its structure. The fixed anions in the pores which are filled with water exert an electrostatic influence on the ions of the electrolyte so that a diffuse electric layer is formed in the pores (see Figure 72). The effective thickness of the diffuse layer strongly depends on the concentration of the electrolyte (see page 123). Thus, at low electrolyte concentrations the pores are filled only by the gegenions (see page 53) which, in the present case, are positively charged. The coions are completely absent from the pores. Only the cations mediate the electric contact between both solutions (this may be directly demonstrated using radioactively labelled ions). At equilibrium the electrochemical potentials of the cations are equal so that for a dilute solution

$$\Delta\varphi_M = -\frac{RT}{F}\ln\left(\frac{c_2}{c_1}\right)$$

This potential difference is termed the Nernst potential, which is closely related to the Nernst equation (page 84). With an anion-exchanging membrane (containing, for example, fixed $-N(CH_3)_3^+$ groups) with otherwise identical properties (average pore thickness, number of ionized groups per unit area of the pore surface) the Nernst potential has the same value but the opposite sign. The membrane that permits only ions of one sign to permeate is called *permselective*.

At higher electrolyte concentrations the effective thickness of the diffuse electric layer becomes comparable with the pore radius and even smaller. The coions penetrate into the pores and the membrane acquires the properties of a liquid junction still influenced by the presence of fixed ions on the walls of the pores. Finally, at rather high electrolyte concentrations, the double layer shrinks to a simple Helmholtz structure (page 123) and the inside of the pore is an electroneutral mixture of gegenions and coions. Thus, the membrane has degenerated to a mere diaphragm with an internal liquid junction with a diffusion potential according to the equation on page 71. By calculation the value of -12 mV is obtained.

J. Bernstein in 1902 and L. Michaelis in 1925 suggested describing the behaviour of this membrane type by analogy with the equation on page 71:

$$\Delta\varphi_M = -\frac{RT}{F}(\tau_B - \tau_A)\ln\left(\frac{c_2}{c_1}\right)$$

where τ_B and τ_A are the transport numbers of the monovalent cation and of the monovalent anion, respectively, which depend on the concentration of the electrolytes in contact with the membrane, on the sign of the fixed ions and on the membrane structure (e.g. for a cation-exchanging membrane and rather dilute solutions, $\tau_B \to 1$ and $\tau_A \to 0$).

The ion-exchanging properties of porous membranes display interesting effects under the flow of an electric current across the membrane. At low concentrations the membrane only permits passage of ions of opposite sign to the fixed ionic groups, which can be utilized in preparative electrolysis. The membrane acts as a separator preventing the ionic products formed at one electrode from reaching the other electrode. At higher concentrations it loses its

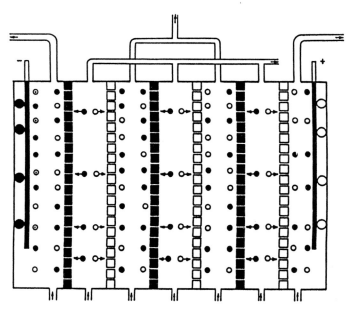

Figure 73. An electrodialytic cell for the desalination of water. The thick line segment with a minus sign is the cathode where hydrogen bubbles (●) are evolved. The anode is represented by the thick line segment with a plus sign. ○ represents a chlorine bubble. Small full circles are sodium ions, small void circles chlorine ions. ■■■■ is the cation exchanging membrane and □□□□ the anion-exchanging membrane. (According to Fischbeck)

discriminating properties but keeps the gases at the electrodes where they should react (see page 97).

Such properties are exhibited by NAFION, a perfluorostyrene sulphonate whose application to brine electrolysis has already been mentioned on page 113.

A very attractive application of membrane electrolysis is *electrodialysis*. By means of dialysis the high- and low-molecular-weight components of a solution are separated in an experimental arrangement where a membrane (not necessarily an electrochemical membrane) is placed between the solution and pure water. Only low-molecular-weight species can permeate across the membrane while the high-molecular-weight components remain in the original solution. Electrodialysis is based on the ability of the porous ion-exchanging membrane to permit permeation of ions of definite charge sign—either cations or anions. The main features of this method will now be shown using the example of desalination of drinking water, which is at present the most important application of electrodialysis. Figure 73 depicts an electrodialytic cell consisting of parallel alternating cation- and anion-exchanging membranes.

Water containing sodium chloride is brought into the cell from below. When an electric current flows across the membrane system one compartment will be enriched with sodium chloride while desalinated water will remain in the other one. Desalinated water is then pumped from the upper part of the cell for direct use while the more concentrated sodium chloride solution is drained for further processing.

References

1. A. H. Bretag (ed.), *Membrane Permeability: Experiments and Models*, Techsearch Inc., Adelaide, 1983.
2. D. S. Flett (ed.), *Ion Exchange Membranes*, Ellis Horwood, Chichester, 1983.
3. P. Meares (ed.), *Membrane Separation Processes*, Elsevier, Amsterdam, 1976.
4. K. S. Spiegler, *Principles of Desalination*, Academic Press, New York, 1969.

ANALYSIS BY MEMBRANE SYSTEMS

Let us prepare a dilute solution of potassium tetraphenylborate (see page 150) in a suitable organic, water-immiscible, solvent (e.g. a mixture of nitroxylenes). This solution is mixed with a PVC solution in cyclohexanone and poured on a polished glass plate with edges to prevent spilling. After evaporation of the solvent, which takes about two days, a small circle of the resulting PVC film is cut and cemented with a PVC solution to a PVC tube. The inside of the tube is filled with 0.1 M KCl with a submerged silver chloride electrode. In this way a potassium ion-selective electrode (ISE) is fabricated. After equilibration in 0.1 M KCl the ISE is immersed in KCl solutions of varying concentrations and the e.m.f. of the cell consisting of this electrode and a reference (e.g. calomel) electrode is determined. The ISE potential (obtained by subtraction of the electrode potential of the reference from the e.m.f.) increases by about 56 mV with a tenfold increase of potassium chloride concentration. This value is close to the Nernst coefficient of 59 mV and the deviation can be explained by a decrease of the activity coefficient of K^+ with increasing concentration. The potassium ISE behaves like a cationic electrode (see page 86).

First the structure of the PVC film ISE will be discussed. The micrograph (Figure 74) shows that in ISEs of this type the islets of the liquid ion exchanger (in the present case, potassium tetraphenylborate dissolved in nitroxylene) communicate with one another in the PVC matrix. On contact of this solution present in the surface of the membrane with the aqueous potassium chloride solution (such an interface of two immiscible electrolyte solutions will be dealt with in more detail on page 149) a Galvani potential difference is formed (cf. page 80):

$$\varphi(\mathrm{w}) - \varphi(\mathrm{o}) \equiv \Delta_\mathrm{o}^\mathrm{w}\varphi = \Delta_\mathrm{o}^\mathrm{w}\varphi_{\mathrm{K}^+}^0 - \frac{RT}{F}\ln\left[\frac{c_{\mathrm{K}^+}(\mathrm{w})}{c_{\mathrm{K}^+}(\mathrm{o})}\right]$$

This equation where activity coefficients were neglected is directly deduced with the help of equality of electrochemical potentials of K^+ in water (w) and in organic solvent (o). It is completely analogous to the Nernst equation (page 84) which of course concerned electron transfer, whereas here ion transfer takes place. Even the standard potential difference $\Delta_\mathrm{o}^\mathrm{w}\varphi_\mathrm{k}^0$, which is characteristic for the ion transferred, is an analogy of the standard electrode potential. On the inner side of the membrane which is in contact with the solution with a

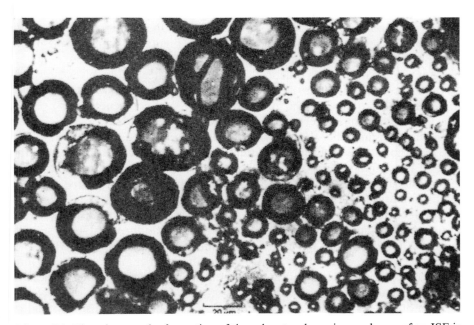

Figure 74. The micrograph of a section of the solvent-polymeric membrane of an ISE in $475 \times$ magnification. (According to G. H. Griffiths, G. J. Moody and J. D. R. Thomas)

potassium concentration of $c'_{K^+}(w)$ the same relationship is valid as above but with an opposite sign. Since the membrane potential is given by the sum of potential differences at the inner and at the outer interface the same Nernst potential as in equation on page 140, is obtained.

However, this remarkable property of potassium ISE has distinct disadvantages. On addition of a sodium chloride solution to the original potassium chloride solution the dependence of the e.m.f. on potassium concentration remains practically the same as in the absence of sodium chloride as long as the ratio of Na^+ to K^+ is less than 5:1. Above that the ISE potential ceases to depend on potassium concentration. The selectivity for K^+ with respect to Na^+ is increased when a more hydrophobic (see page 151) anion than tetraphenyl-borate, such as p-chlorotetraphenylborate

is used as the membrane electrolyte (liquid ion-exchanger). In this case, the potassium ISE ceases to respond to potassium ions only at a ratio of $Na^+:K^+$ above 20. A still larger increase of selectivity is achieved if potassium ions are bound in a complex, present, of course, only in the membrane. A suitable complexing agent is the antibiotic valinomycin (Figure 75) which is one of the ionophores (ion carriers) whose properties will be discussed in more detail on page 163. This substance is practically insoluble in water but dissolves quite well in organic solvents; it binds the potassium ion very strongly in its internal cavity while its interaction with Na^+ is much weaker. Valinomycin–potassium ISE reacts to potassium ions even in a thousandfold excess of Na^+!

Sodium ions influence the behaviour of the potassium ISE by penetrating into the membrane and displacing potassium ions into the aqueous solution. In the case of the valinomycin–potassium ISE this exchange reaction becomes much more difficult as the sodium ions have to remove the rather strongly bound potassium ions from its complex.

The PVC film ISE, often called solvent–polymeric ISE, is the most important member of liquid-membrane ISEs, which are outstanding analytical sensors not only for potassium but also for calcium, lithium, sodium, acetylcholine, nitrate, chloride and other ions. Most often other ion-exchanger ions besides p-chlorotetraphenylborate and other complex-forming agents besides valinomycin are used in these ISEs.

In another version of ISEs solid-electrolyte membranes are used instead of liquid membranes. The membrane materials are, for example, LaF_3 (fluoride ISE), AgI (iodide ISE), $AgCl$ (chloride ISE) and Ag_2S (sulphide ISE).

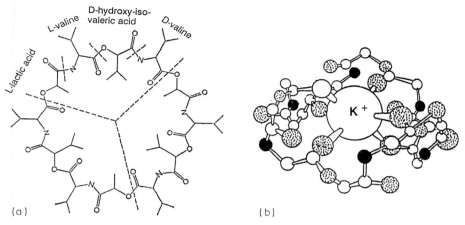

Figure 75. Structure of (a) valinomycin and of (b) its potassium complex. (According to W. L. Duax and coworkers)

The oldest and most often used ISE is the *glass electrode*. In 1906 the German botanist M. Cremer used a thin glass shell as a biological membrane model. Later it was shown that the membrane potential of this ISE depended on pH according to a simple equation (at 25 °C)

$$\Delta\varphi_M = \text{constant} - 0.059\,\text{pH}$$

The resistance of the glass electrode is high and this is the reason why it only became a favourite laboratory tool after the US company Beckman introduced an electronic potentiometer (pH-meter) on the market in 1936. The glass usually employed for the preparation of glass electrodes consists of 22% Na_2O, 6% CaO and 72% SiO_2. The behaviour of the glass electrode is affected when pH is increased towards the alkaline pH range in solutions containing sodium ions. Finally, at pH > 11 the membrane potential ceases to respond to hydrogen ions (some special glasses also containing Li_2O and BaO admixtures have a higher limit for this critical pH). This 'sodium error' is connected with the mechanism of formation of the membrane potential. In contact with water a hydrated layer approximately 100 nm thick is formed on the glass surface (therefore the glass electrode should be first soaked in buffer before use). Sodium ions from this layer are exchanged for hydrogen ions from the solution,

$$H_3O^+(\text{solution}) + Na^+(\text{glass}) = H^+(\text{glass}) + Na^+(\text{solution}) + H_2O$$

At pH < 11 this exchange equilibrium is shifted to the right and practically all exchangeable sodium ions in the hydrated layer are replaced with hydrogen ions. The membrane potential obviously follows the equation on page 143, where c_{K^+}'s are substituted with c_{H^+}'s. As at pH < 11 this concentration is practically constant the membrane potential only depends on pH as shown above. Another situation occurs at pH > 11 and the exchange equilibrium shifts to the left. Now simultaneously with an increase of pH the H^+ concentration in the glass is diminished so that the membrane potential no longer depends on pH. The result is shown in Figure 76. This property is enhanced if a part of silicon oxide is replaced with aluminium oxide. Such glass electrodes then function as sodium ISEs.

Glass electrodes as well as other ISEs have a large scope of applications. The glass electrode is nearly the exclusive sensor for pH determination in a great variety of media (for an exception, see page 49). The degree of contamination of water is determined with fluoride and chloride ISEs. Automatic devices for serum and blood analysis predominantly involve ISEs for the determination of K^+, Na^+ an Cl^-. In electrophysiology ion-selective microelectrodes have found wide application. They are basically micropipettes (see page 87), whose tiny tip is filled with an ion-exchange solution. A liquid bridge connecting a reference electrode is sealed to the micropipette (Figure 77). In the case of measurement in excitable tissues like nerves or muscles this arrangement obviates the effects of the omnipresent electric field. When measuring in single cells it is necessary to pierce the cell membrane (this has no profound effect on its behaviour as the diameter of the tip is $1-3$ μm only). In this way intracellular concentrations of potassium, sodium, chloride and calcium ions are determined while glass microelectrodes are used for intracellular pH measurement. In a similar way, ion concentrations in the intercellular liquid are analysed.

Devices based on the glass electrode can be used to determine certain gases present in a gaseous or liquid phase. Such a gas probe (Figure 78) consists of a glass electrode covered by a thin film made of a plastic material with very small pores, which is hydrophobic, so that the solution cannot penetrate into the pores

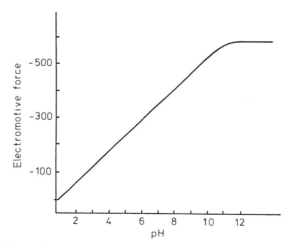

Figure 76. The dependence of the e.m.f. of the glass electrode/saturated calomel electrode cell on the pH of the analysed solution

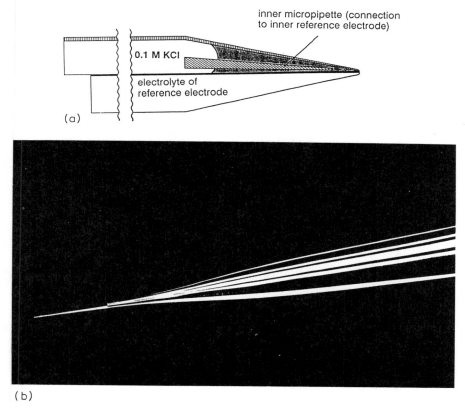

(a)

(b)

Figure 77. (a) A potassium ion-selective microelectrode of the double-barrel type. (According to J. L. Walker). (b) A micrograph of the capillary tip of a double-barrel ammonium-selective microelectrode (by permission of D. de Beer)

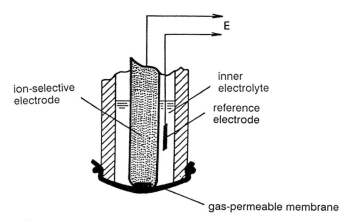

Figure 78. A gas probe. (Manufactured by Orion Research)

148

(cf. page 130). A thin layer of an indifferent solution is present between the surface of the film and of the glass electrode and is in contact with a reference electrode. The gas (ammonia, carbon dioxide, etc.) permeates through the pores of the film and dissolves in the solution at the surface of the glass electrode. A definite value of the pH corresponding to the equilibrium concentration of a gas such as ammonia fixes the potential of the glass electrode in this solution. The actual determination of the ammonia concentration is based, of course, on a calibration plot of the electrode potentials versus the known ammonia concentration in the solution analysed.

References

1. G. Eisenman (ed.), *Glass Electrodes for Hydrogen and Other Cations: Principles and Practice*, Marcel Dekker, New York, 1967.
2. J. Janata, *Principles of Chemical Sensors*, Plenum Press, New York, 1989.
3. J. Koryta and K. Štulík, *Ion-Selective Electrodes*, 2nd ed., Cambridge University Press, Cambridge, 1983.
4. W. E. Morf, *The Principles of Ion-Selective Electrodes and of Membrane Transport*, Akadémiai Kiadó, Budapest, and Elsevier, Amsterdam, 1981.
5. E. Syková, P. Hník and L. Vyklický (eds.), *Ion-Selective Microelectrodes and Their Use in Excitable Tissues*, Plenum Press, New York, 1981.
6. A. P. F. Turner, I. Karube and G. S. Wilson, *Biosensors, Fundamentals and Applications*, Oxford University Press, Oxford, 1989.
7. I. Zeuthen (ed.), *The Application of Ion-Selective Microelectrodes*, Elsevier, Amsterdam, 1981.

BIOSENSORS

Composite systems also comprise the *enzyme electrodes*. The basic component of these sensors is a hydrophilic polymer containing an enzyme which transforms the determinand into a substance that is sensed by the electrode. For example, the urea electrode consists of an ammonia probe (Figure 78) where the porous film is covered with a nylon net (G. G. Guilbault who invented this sensor originally used a piece of nylon stocking). A polyacrylate gel containing the enzyme urease is spread on this net. Urea penetrates from the analysed solution into the gel layer where it is decomposed according to the equation

$$CO(NH_2)_2 + H_2O \xrightarrow{\text{urease}} CO_2 + 2\,NH_3$$

Instead of potentiometric indication, some sensors are based on amperometric determination of the product produced by the enzymatic reactions. For example, an electrochemical sensor for determination of glucose in blood contains the glucose oxidase immobilized in polyacrylamide gel and the Clark oxygen sensor (page 130). The enzyme reaction is

$$\text{Glucose} + O_2 \xrightarrow[\text{oxidase}]{\text{glucose}} H_2O_2 + \text{gluconic acid}$$

The decrease of oxygen reduction current measured with the Clark sensor indicates the concentration of glucose.

While enzyme electrodes (the glucose sensor being the best known one) find their use in clinical laboratories only with some difficulty this is even more true for *bacterial* or *tissue electrodes*. Attempts have been made for a number of years to substitute the immobilized enzyme layer in an enzyme electrode with a suspension of bacteria pressed to the surface of a glass electrode or with a tissue section. Such systems can perhaps be used in a specialized laboratory where an enthusiastic and experienced worker personally carries out the analyses, but the difficulty of proper standardization prevents these methods from being used in routine analysis. Unwarranted hopes stem as usual from frequent reviews in popular journals (the 'biochips' being their favourite subject).

References

1. Page 148, Ref. 6.
2. F. Scheller and F. Schubert, *Biosensoren*, Akademie-Verlag, Berlin, 1989.

ELECTROCHEMICAL EXTRACTION

More details on the electric properties of the *interface of two immiscible electrolyte solutions* (ITIES) will give a better understanding of the behaviour of ISEs, clearer discrimination between the terms, hydrophobic and hydrophilic (page 29), etc. Two liquids with low miscibility, for example water and tetrachloromethane, CCl_4, will be examined. The mutual solubility of these liquids is very low. Iodine is dissolved in CCl_4 to give a concentration of 10^{-3} mol dm^{-3}. When the resulting solution is shaken with the same volume of water some of the iodine enters the aqueous phase. Analysis of the resulting content of iodine in the two solvents reveals that the ratio of its concentrations in tetrachloromethane and water is 85 at 25°C. If the initial concentration of iodine in CCl_4 is increased to 10^{-2} mol dm^{-3}, the resultant concentration ratio after shaking with water is the same. The distribution equilibrium of a non-ionized substance between water (w) and CCl_4 (o) is characterized by the equilibrium constant, called the distribution (partition) coefficient

$$K_{I_2}^{o,w} = \frac{[I_2]_o}{[I_2]_w}$$

The only cause of unequal distribution of iodine between the two solvents is the much stronger solvation in the organic solvent. Distribution of an electrolyte between those two solvents will now be discussed. On agitating an aqueous solution of tetrabutylammonium iodide with CCl_4 we find that the experiment was not chosen favourably. The tetrabutylammonium iodide passes almost completely into the non-aqueous phase. Nonetheless, the conductivity of this phase is negligible. The iodide anions and tetrabutylammonium cations are completely associated in ion-pairs (because of the low permittivity of the solvent, cf. pages 37 and 44). Consequently, another non-aqueous solvent that is immiscible with water must be chosen where the salt is soluble and, at the same time, almost completely dissociated.

This condition is fulfilled for millimolar concentrations of salts in solvents such as nitrobenzene with a relative permittivity above 30. The equation for the distribution coefficient can be written in two ways:

$$K_{BA}^{o,w} = \frac{c_{BA,o}}{c_{BA,w}} = \sqrt{\frac{[B^+]_o[A^-]_o}{[B^+]_w[A^-]_w}}$$

where the c_{BA} values are the overall concentrations of BA, the tetrabutylammonium iodide. Contribution of the cation and of the anion to partition between the two phases is doubtless different and, therefore, a characterization with the help of individual ion partition coefficients may be attempted. They are defined as $k_{B^+} = [B^+]_o/[B^+]_w$ and $k_{A^-} = [A^-]_o/[A^-]_w$, which when combined give the partition coefficient of the salt:

$$K^{o,w} = (k_{B^+} \, k_{A^-})^{1/2}$$

Because dividing $K_{BA}^{o,w}$ into k_{B^+} and k_{A^-} is rather arbitrary, defining one of the ionic species as hydrophobic and the other as hydrophilic will be of little value. Thus, a system of reference is needed for which more or less reliable values of k_{B^+} and k_{A^-} are known. As the distribution coefficient characterizes different solvation of particles in two solvents the cation and the anion of such a salt must be almost equally solvated in various solvents. A. J. Parker recommended tetraphenylarsonium tetraphenylborate (TPAs TPB) as a salt of choice:

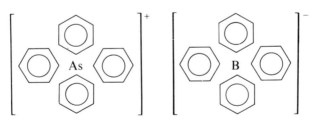

Both of the ions are voluminous so that the central atom only acts electrostatically on solvent dipoles. The interaction of phenyl groups and the solvent is practically the same with both of the ions.

Thus we may be justified in assuming equal individual partition coefficients of TPAs$^+$ and TPB$^-$ for any pair of solvents. To obtain the value of the individual partition coefficient between two solvents of interest for an arbitrary ion, e.g. a cation, first the distribution coefficient for the salt TPAs TPB is determined and hence the individual partition coefficient of TPAs$^+$ and TPB$^-$, which are, of course, equal. Then the distribution coefficient of the salt of that cation and TPB$^-$ is found and, using the above equation, also k_{B^+}. The values of individual partition coefficients are listed in Table 9.

Table 9. Individual distribution coefficients $k^{o,w}$ and standard potential differences $\Delta_{nb}^{w} \varphi_i^0$ in the system water (w)/nitrobenzene (nb) for individual ionic species i

Ion	$k_i^{o,w}$	$\Delta_{nb}^{w} \varphi_i^0 / V$
Li^+	2.6×10^{-7}	0.389
Na^+	8.6×10^{-7}	0.358
Ca^{2+}	10^{-6}	0.354
H^+	1.9×10^{-6}	0.337
K^+	5.4×10^{-5}	0.252
Rb^+	3.2×10^{-4}	0.206
Cs^+	1.9×10^{-3}	0.161
Me_4N^+	0.26	0.035
Bu_4N^+	1.8×10^4	-0.251
Ph_4As^+	2.4×10^6	-0.372
Cl^-	4.4×10^{-6}	-0.316
Br^-	10^{-5}	-0.295
NO_3^-	5.2×10^{-5}	-0.253
I^-	5×10^{-4}	-0.195
Picrate	6.5	0.048
Tetraphenylborate	2×10^6	0.372

The individual ion partition coefficients are linked to standard potential differences between two solvents for a given ion (page 143) by equations

$$\Delta_o^w \varphi_{B^+}^0 = -\frac{RT}{F} \ln k_{B^+}^{o,w}$$

$$\Delta_o^w \varphi_{A^-}^0 = \frac{RT}{F} \ln k_{A^-}^{o,w}$$

These constants are also listed in Table 9.

Individual ion partition coefficients, or standard potential differences, respectively, will help to classify hydrophobic and hydrophilic ions. Nitrobenzene will be chosen as the solvent of reference. Ions whose individual partition coefficient lies around unity are designated as semi-hydrophobic while ions with $k^{o,w} \gg 1$ as hydrophobic and those with $k^{o,w} \ll 1$ as hydrophilic.

Let us consider a salt of a hydrophobic cation and of a hydrophilic anion distributed between water and organic solvent. The cation has the 'tendency' to cross the boundary (ITIES; see above) while the tendency of the anion is just the opposite. These transitions, at least to the bulk of solution, are indeed not possible but the charge distribution occurs at ITIES by formation of an electric double layer charged positively in the organic solvent and negatively in water. More detailed investigations have shown that it only consists of two diffuse electric layers (see page 123). The charge distribution results in the formation of an electric potential difference between the two media which is called the *distribution potential*. In the simple case of an electrolyte composed of univalent cations and anions a straightforward manipulation of the equation on page 143 for the cation and of an analogous equation for the anion gives the relationship for the distribution potential:

$$\Delta_o^w \varphi_{distr} = \frac{\Delta_o^w \varphi_{B^+}^0 + \Delta_o^w \varphi_{A^-}^0}{2}$$

The ITIES has further interesting properties, particularly when one of the solvents is water and the other is an organic liquid and when the aqueous phase contains a strongly hydrophilic electrolyte (e.g. lithium chloride) and the organic phase a strongly hydrophobic electrolyte like tetrabutylammonium tetraphenylborate. Then the properties of the ITIES become completely analogous to those of a polarized electrode (see page 122).

The potential difference across the ITIES can be changed by introducing charges of opposite sign into each phase using electrodes immersed into these phases, most conveniently with a four-electrode potentiostatic system (for a three-electrode potentiostatic arrangement, see page 99), as shown in Figure 79. When, for example, a semi-hydrophobic cation is present in the aqueous phase it can be 'forced' to cross the ITIES and to enter the organic phase. This process is

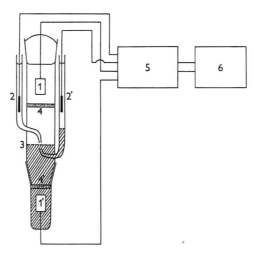

Figure 79. A device (voltage clamp) for polarization of the interface of two immiscible electrolyte solutions (ITIES) by means of a four-electrode system. Electric current is fed into the system of two immiscible liquids (the upper solution 4 is an aqueous electrolyte, the lower 4′ is a solution in an organic solvent) by means of two auxiliary platinum electrodes, 1 and 1′. The reference electrodes 2 and 2′ are connected by means of Luggin's capillaries to both solutions. The dashed line 3 is the actual interface (ITIES). A four-electrode potentiostat together with a pulse generator 5 and the recorder 6 are connected to the electrode system

connected with the flow of an electrical current, where the strongly hydrophilic electrolyte in the aqueous phase and the strongly hydrophobic electrolyte in the organic phase act as indifferent electrolytes (cf. page 72). The value of $\Delta_o^w \varphi$ necessary for the transfer of a particular ion across the ITIES is connected with the value of the individual partition coefficient for that ion. This situation is completely analogous to electrolysis with metallic electrodes and the same methods of investigation as described on pages 100 and 127 can be applied to electrolysis at ITIES.

References

1. J. Koryta, 'Electrochemical polarisation of the interface of two immiscible electrolyte solutions', *Electrochim. Acta*, **24**, 293 (1979); **29**, 445 (1984); **33**, 189 (1988).
2. A. J. Parker, 'Solvation of ions—enthalpies, entropies and free energies of transfer', *Electrochim. Acta*, **21**, 671 (1976).
3. P. Vanýsek, *Electrochemistry at Liquid–Liquid Interfaces*, Springer-Verlag, Berlin, 1985.

DOWN TO THE MOLECULAR LEVEL—CELL WALLS AND CELL MEMBRANES

The scale of our view of membranes will be drastically changed now as from previously looking at systems of a tenth of a millimetre or more thick it will shift to layers of tenths of a micrometre or less thick. As already mentioned (page 137), a real biological membrane can only be seen on micrographs recorded by means of an electron microscope. In an optical microscope a cell wall may be seen that is typical of cells of plants, algae, fungi and bacteria. The system cell wall—cell or plasma membrane differs from one class of organism to the other and is sometimes rather involved.

The cell wall of *green plants* is built from polysaccharides, such as cellulose or hemicelluloses (water-insoluble polysaccharides usually with a branched structure), and occasionally from a small amount of glycoproteins (see pages 154 and 177). Carboxyl groups of the bound glucuronic acid invest the plant cell wall with slightly acid properties. In this case the cell wall is a porous structure with fixed charges on the surface of the pores, thus resembling an ion-exchange membrane (page 139):

Glucuronic acid

Gram-positive bacteria (which stain blue in the procedure suggested by H. Ch. J. Gram in 1884) have their cell wall built of cross-linked polymers of amino acids and sugars (peptidoglycans; see Figure 80c). The actual surface of the

154

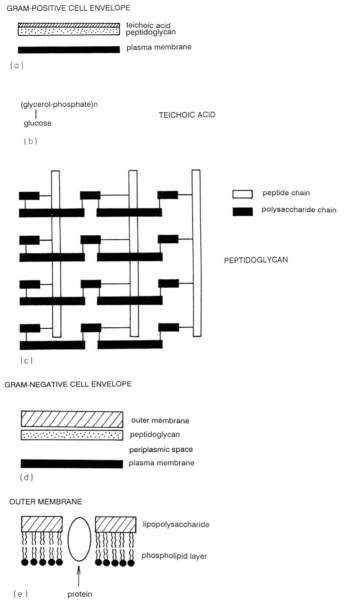

Figure 80. Cell envelopes of bacteria: (a) a gram-positive bacterium with (b) teichoic acid and (c) a glycoprotein, peptidoglycan, as structural units. (d) A gram-negative bacterium envelope showing (e) its outer membrane structure. The basic unit of a glycolipid (glycopolysaccharide) is glucosamine with attached long alkyl-chain fatty acids (inner hydrophobic part) and a long sugar chain with sparsely bound phosphate groups (outer hydrophilic part). The periplasmatic space contains proteins that transport sugars and other nutrients

bacterium is formed by teichoic acid which makes it hydrophilic and negatively charged. In this case the cell wall is a sort of giant macromolecule—a bag enclosing the whole cell. Penicillin inhibits the synthesis of this cell wall.

The surface structure of *gram-negative bacteria* (these are not stained by Gram's method and must be additionally stained red with carbol fuchsin) is more diversified (Figure 80d). It consists of an outer membrane whose main building unit is a lipopolysaccharide together with phospholipids and proteins (Figure 80e). In this membrane, protein channels are present, the so-called porins, which permit hydrophilic, low-molecular-weight substances to pass across the membrane. The actual cell wall is made of a peptidoglycan but the destructive effect of penicillin is prevented here by the outer membrane.

The seemingly bare (actually enclosed by a real plasma membrane) animal cells possess a sort of loosely bound envelope on their surface which is built of polysaccharides (glycocalyx).

References

1. I. Darnell, H. Lodish and D. Baltimore, *Molecular Cell Biology*, W. H. Freeman & Co., New York, 1986.
2. J. B. Finean, R. Coleman and R. H. Michell, *Membranes and Their Cellular Function*, Blackwell, Oxford, and Halsted Press, New York, 1979.
3. A. Kotyk, K. Janáček and J. Koryta, *Biophysical Chemistry of Membrane Transport*, John Wiley & Sons, Chichester, 1989.
4. L. Stryer, *Biochemistry*, W. H. Freeman & Co., New York, 1984.

BIOLOGICAL MEMBRANES: THEIR STRUCTURE AND MODELS

Why are cells and organelles enclosed by membranes at all? The answer is simple— without them basic life functions could not proceed. Membranes are a defined boundary against the cell environment. They control the volume of the cell, internal pH and ionic composition. They permit extrusion and accumulation of substances acting as sources of energy (metabolic fuels). They mediate the movement of parts of the cell (e.g. cilia and flagella) as well as transport of substances along the cell surface. They are the sites for energy flow based on transduction of one form of energy to another and for information flow by means of physical or chemical signals. Various catalytic reactions proceed on membranes. Indeed, life could not exist without membranes.

The actual plasma membrane is very thin—between 5 and 10 nanometres thick. It consists, like the whole organism, of lipids, proteins and sugars which are present either separately or in the form of lipoproteins, glycolipids and glycoproteins. Basic structure types of biomembranes are shown in Figure 81, from which it is obvious that the representation of various components markedly differs from one type to another one. The weight proportions of lipids and proteins in typical cell membranes is shown in Figure 82.

156

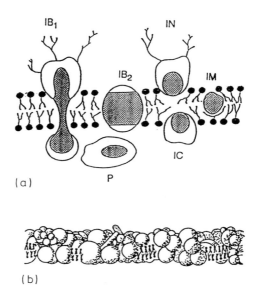

(a)

(b)

Figure 81. Typical biological membrane structures. (a) A liquid-mosaic model in the form proposed by A. Kotyk with different types of disposition of membrane proteins (phospholipids are shown as black circles with two waved tails). IB_1—integral, membrane-bridging protein, with a single polypeptide span; IB_2—the same with several spans; IN and IC—integral non-cytoplasmic and cytoplasmic proteins; IM—integral, buried proteins; P—peripheral protein. (b) A model of the inner membrane of mitochondrion where the proteins (rods and balls) strongly prevail over the phospholipids. (According to A. Patel)

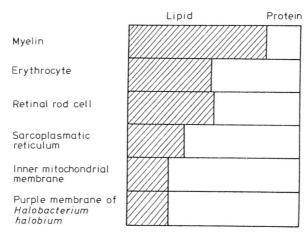

Figure 82. The lipid–protein ratio in various biological membranes. (According to J. B. Finean, R. Coleman and R. H. Mitchell)

Sugars represent 1–8% of the weight of a dried membrane of mammalian muscle whereas 25% of that of an amoeba membrane. The main sugar components are L-fucose, D-galactose, D-mannose, sometimes D-glucose, N-acetyl-D-galactosamine and N-acetyl-D-glucosamine. Important components are sialic acids which are sugars with acid properties functioning as terminal units in glycoprotein or glycolipid chains. N-Acetylneuraminate is an anion of one of the sialic acids:

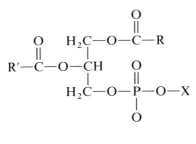

In its ionized form it markedly contributes to the surface charge of the membrane.

At least with some membranes the basic structural matrix is the bimolecular lipid layer, the 'bimolecular leaflet'. The lipids are oriented with their hydrophobic alkyl tails towards the inside of the membrane while their polar heads are directed towards the adjacent solution (Figure 83).

The lipidic component of a membrane is represented primarily by phospholipids. The most important is the group of phosphoglycerides. The basic structure is phosphatidic acid:

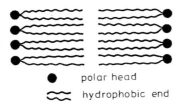

Figure 83. A scheme of the bilayer lipid membrane (BLM). The black circles indicate the polar heads (the hydrophilic part) consisting of phosphoric acid and ethanol amine and analogue derivatives, the waved lines are the long alkyl chains of fatty acids (the hydrophobic part)

where in the formula $X = H$, and R and R′ are alkyl or alkenyl long-chain groups. Thus, glycerol is esterified with two higher fatty acids, such as stearic, palmitic, oleic, linoleic, etc., acids, and the orthophosphoric acid is bound on the remaining OH group. One of its acidic groups is esterified with ethanolamines, usually substituted on the nitrogen atom. For example, with binding choline $(X = CH_2CH_2N(CH_3)_3{}^+)$ the phospholipid lecithin is formed. In place of ethanolamines, other molecules such as serine, inositol, etc., can be bound to the phosphoric acid.

Instead of glycerol some phospholipids contain the ceramide,

$$\begin{array}{l} CH(OH)CH{=}CH(CH_2)_{12}CH_3 \\ | \\ CH{-}NHOCR \\ | \\ CH_2OH \end{array}$$

where R is a long-chain alkyl group. The ester of ceramide with phosphoric acid, also esterified with choline, is sphingomyeline. When ceramide is bound to a sugar, such as glucose or galactose through the β-glycosidic bond, cerebrosides are formed. Lipids also include cholesterol. In Table 10 the presence of individual types of lipids in various biological membranes is demonstrated.

Proteins are bound to a membrane either only loosely by electrostatic forces (extrinsic proteins) or by hydrophobic interactions. In the latter case they are immersed to differing depths of the membrane (integral proteins; Figure 81a), most of them actually spanning the membrane. This sort of protein either reinforces membrane structure or exhibits various transport (see pages 168 and 188) or catalytic functions. Proteins in membranes of mitochondria and thylakoids are arrayed in protein complexes.

Table 10. Representation of lipids (in molar%) in biological membranes. (According to A. Kotyk)

Membrane type	PC	PE	PS	PI	PG	CL	Sphingolipids	Sterols	Others
Cytoplasmic membrane									
of bacterium *Salmonella typhimurium*	—	81	—	—	32	7	—	—	1
of bacterium *Bacillus subtilis*	—	—	—	—	78	3	—	—	19
Plasma membrane									
of rat liver cell	28	17	6	6	—	1	12	27	3
Myelin sheath									
of human nerve	11	14	5	1	—	—	32	25	12
Inner membrane of mitochondrion									
from rat liver	40	35	1	5	2	17	1	1	1
from cauliflower	37	34	—	4	3	13	—	5	4

PC = phosphatidylcholine (lecithin), PE = phosphatidylethanolamine (cephalin), PS = phosphatidylserine, PI = phosphatidylinositol, PG = phosphatidylglycerol, CL = diphosphatidylglycerol (cardiolipin).

Overlapping of various physical and chemical processes in living organisms makes it extraordinarily difficult to investigate the separate steps of those processes. Their disentangling is helped by various measures aimed at maximum simplification. For example, methods of genetic engineering hereditarily remove some functions of the organism (usually of bacteria). Membrane processes are studied with the help of membrane models with built-in suitable isolated components of a biological system. The most favourite model of biological membranes is the *bilayer lipid membrane* (BLM).

A cell containing a dilute electrolyte is divided into two compartments by a Teflon septum with a small hole (diameter about 1 mm). When a drop of a lipid in a suitable solvent, e.g. octane, is placed on the hole, a striking phenomenon is observed. The layer of the lipid solution gradually becomes thinner, rainbow interference colours appear on it and later also black spots and, finally, the whole layer becomes completely black. The blackening marks the transition of the lipid layer from a multimolecular thickness to a bimolecular film, the bilayer lipid membrane (BLM) which is a non-reflecting optically black region. A BLM in a solution was first prepared by P. Mueller, D. O. Rudin, H. Ti Tien and W. C. Westcott in 1962, although similar phenomena with soap bubbles had already been observed by R. Hooke in 1672 and I. Newton in 1702. The attenuation of the layer is caused by interfacial forces together with the dissolution of the membrane solvent in the adjacent aqueous solution. The thick edge of the membrane is called the Plateau–Gibbs boundary (Figure 84). The thickness of the membrane has been determined by means of low-angle light reflection,

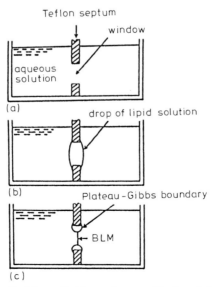

Figure 84. Preparation of a bilayer lipid membrane with steps (a), (b) and (c) described in the text. (According to P. Mueller and coworkers)

160

electric capacity measurement and electron microscopy (a metal coating can, of course, result in artefacts).

Another method of preparation of BLM is demonstrated in Figure 85. A drop of phospholipid solution is attached to the tip of a glass capillary and sucked inside. The thickness of the BLM obtained by these two methods varies between 5 and 8 nanometres. Originally, phospholipids isolated, for example, from egg yolk, erythrocytes, brain tissue, etc., were used as materials of BLMs. However, in order to exclude the influence of various admixtures (which, by the way, often increased the stability of the BLM) synthetic phospholipids are now mainly used.

The two long alkyl chains present in the phospholipid molecule are the cause of formation of the BLM instead of micelles which are typical of surfactants with one hydrophobic group in the molecule. Interaction of opposite pairs of these chains is, because of the hydrophobic effect, more advantageous than interaction in a monolayer built of molecules containing such pairs (cf. page 29).

A certain analogy of a micelle, but with a bilayer, is the *liposome* (lipid vesicle). It is a spherical structure, usually of a diameter of several tens of nanometres, formed, for example, by ultrasonification of an aqueous suspension of phospholipids (see Figure 86). When a colloidal solution of liposomes with a definite diameter is required it can be obtained from the original suspension by gel filtration.

While planar BLMs are rather easily investigated by electrochemical methods this is not possible in the case of such tiny bodies as liposomes. The membrane potential is measured only indirectly by means of a membrane-soluble ion (tetraphenylphosphonium cation, for example). At equilibrium this ion is distributed between the bathing solution and the inside of the liposome according to the Nernst potential. Since the total concentration of this 'pilot' ion and its concentration in the bathing solution can easily be determined, the

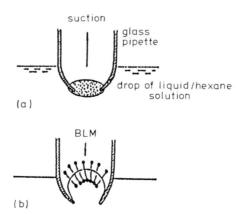

suction

glass pipette

drop of liquid/hexane solution

(a)

BLM

(b)

Figure 85. Preparation of the BLM by sucking of a phospholipid solution into a glass capillary

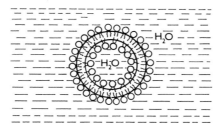

Figure 86. A liposome

membrane potential is obtained from the Nernst potential equation (page 140). A planar BLM cannot be investigated by molecular spectroscopy because of the small amount of substance in an individual BLM. This disadvantage is removed with liposomes because their suspensions can be quite concentrated. Thus, for application of e.s.r. (= electron spin resonance, the spectra of which indicate the presence and properties of a substance with an unpaired electron), 'spin-labelled' substances are embedded in the liposome membrane, such as a phospholipid with 2,2,6,6-tetramethylpiperidine-N-oxide (TEMPO):

$$\begin{array}{c} N \longrightarrow O \end{array}$$

bound to a long-chain alkyl group of the phospholipid. The properties of the spectrum signal of this substance influenced by the environment of the membrane supply information on the movement and position of the spin-labelled lipid in the membrane.

Liposomes also have important practical applications. They make it possible, for example, to effectively administer medicines enclosed in the inner region of the vesicle, to transfer definite parts of DNA (plasmids) into the cell, etc.

References

1. G. Ceve and D. Marsh, *Phospholipid Bilayers*, John Wiley & Sons, New York, 1987.
2. D. Chapman (ed.), *Biomembrane Structure and Function*, Vol. 4 of Topics in Molecular and Structural Biology, Verlag Chemie, Weinheim, 1984.
3. F. Conti and W. E. Blumberg (eds.), *Physical Methods in Biological Membranes and Their Model Systems*, Plenum Press, New York, 1985.
4. G. Gregoriadis (ed.), *Liposome Technology*, Vols. 1 to 3, CRC Press, Boca Raton, 1984.
5. M. K. Jain, *Introduction to Biological Membranes*, John Wiley & Sons, New York, 1988.
6. C. G. Knight (ed.), *Liposomes: From Physical Structure to Therapeutic Applications*, Elsevier–North Holland, Amsterdam, 1981.
7. A. N. Martonosi (ed.), *Membranes and Transport*, Vols. 1 and 2, Plenum Press, New York, 1982.
8. H-Ti Tien, *Bilayer Lipid Membrane (BLM)*, Marcel Dekker, New York, 1974.

PASSIVE TRANSPORT

Although the BLM is extremely thin, it exhibits enormous resistance at zero membrane potential (between 10^8 and 10^{10} Ω cm^{-2}). The slight conductivity of such a simple 'unmodified' membrane results, perhaps, from the pores which are formed by the thermal motion of the molecules in the membrane. Water molecules enter these 'statistical' pores for a moment and enable the ions to penetrate through the membrane in a short time. When the membrane potential is increased from an external voltage source the frequency and size of the pores grow and, finally, at a potential difference of several hundreds of millivolts the membrane breaks down.

The BLM conductivity increases when certain substances are added to the bathing solution or to the membrane alone. Most of these substances act on biological membranes in a similar way. The BLM is then an adequate model of a naturally occurring membrane. In this group belong, for example, hydrophobic ('lipophilic' = fat-liking) ions which alone are able to penetrate the membrane. Other substances cross the membrane through different mechanisms.

Figure 87 demonstrates several modes of electric charge transfer across the membrane. In addition to the simple transport of a hydrophobic ion, several transport mechanisms through ion carriers (ionophores), channels or pores are shown. These transport paths sometimes require an interaction between the carrier or the channel-forming molecules and the transported ion. This interaction often considerably depends on the individual properties of the ion,

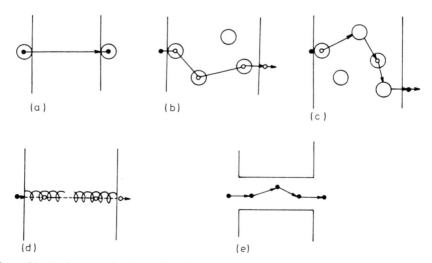

Figure 87. Various mechanisms of passive ion transport across a membrane: (a) transfer of a lipophilic ion; (b) simple carrier transport: the complex of the ion with the carrier migrate without a change across the membrane; (c) carrier relay: the ion is exchanged between different carrier molecules when passing across the membrane; (d) an ion-selective channel; (e) a membrane pore which does not select between the transferred ions

particularly on its size, and is, therefore, nearly always ion-selective. The rather narrow channel (formed as a rule by a molecule with helical shape) is not identical with a pore, which has such a large diameter that ions and solvent molecules can penetrate it regardless of their size.

Hydrophobic (lipophilic) ions which easily penetrate the membrane and considerably increase its conductivity include tetraphenylborate (see page 150), picrate

$$NO_2\text{-}\underset{NO_2}{\overset{NO_2}{C_6H_2}}\text{-}O^-$$

and dipicrylaminate anions

$$\left[NO_2\text{-}\underset{NO_2}{\overset{NO_2}{C_6H_2}}\text{-}N\text{-}\underset{NO_2}{\overset{O_2N}{C_6H_2}}\text{-}NO_2 \right]^-$$

the long alkyl-chain tetraalkylammonium cations, etc. This group of substances includes some anaesthetics with similar structure.

For low concentrations, the membrane conductivity is directly proportional to the concentration of the lipophilic ion. When an electric potential difference is formed between the two bathing solutions, no rectilinear current-potential dependence is obtained. Such a dependence corresponding to Ohm's law would result in the membrane behaving as a simple 'passive' resistance. In fact, the actual current-potential dependence is similar to the diagram in Figure 50(a). The laws governing the ion transfer across the BLM and the electrode reaction rate are obviously analogous.

The carrier mechanism (usually a carrier relay; see Figure 87c) is connected with a group of substances discovered during the last forty years. The structure of envelope helps to solubilize the complex in various organic low-polarity solvents where the free cations cannot be dissolved.

A typical substance in this readily growing family of natural and synthetic ionosphores is valinomycin (cf. page 145) which is synthesized by the primitive fungus *Streptomyces fulvissimus*. Valinomycin is a cyclic depsipeptide with a 36-membered ring; its building units are alternatively amino acids and α-hydroxy acids. Figure 75 shows the bracelet structure of free valinomycin and the cylindrical structure of its potassium complex. The valinomycin network completely encircles the central cation. The complex with its lipophilic exterior can be dissolved in the membranes of cells and cell organelles and, in this way, can destroy an important metabolic function—oxidative phosphorylation (see page 189). This is the basis of *Streptomyces* antibiotic function. All processes that occur with the participation of valinomycin and related substances are ion-selective with respect to alkali metal cations, having different intensities in the presence of various ions. This ion-selectivity is linked to the complex stability, which is strikingly different for individual types of ions. When the alkali metal ion is small, such as the bare lithium ion, it easily fits into the internal cavity of valinomycin. On the other hand, a great deal of energy is needed to liberate this ion from its large hydration sheath, resulting in negligible stability of the valinomycin complex of lithium. The potassium ion has a larger ion radius than the lithium ion but still fits into the cavity, so that the solvation is much weaker than for lithium and, therefore, the tendency to complex formation is much larger. The caesium ion has virtually no hydration sheath but the radius of the free ion is larger than that of the cavity, so that the macrocyclic structure is strained and the binding in the complex is less advantageous than with potassium. Consequently, the caesium complex of valinomycin is less stable than the potassium complex. A completely analogous dependence on the ion radius, as demonstrated by the stabilities of complexes, is exhibited by the resistances of the BLM and by the membrane potentials in the presence of valinomycin and the salts of different alkali metal cations.

Above 30–40°C, the BLM behaves like a liquid with oriented molecules. This state of aggregation is termed liquid-crystalline. When the temperature falls below a critical value characteristic for any BLM material, the aggregation of the membrane becomes crystalline (gelation). Then the motion of both the molecules that form the membrane or that are dissolved in it is considerably restricted. In the presence of valinomycin or other ionophores the conductivity of the membrane strongly decreases as the ionophore loses its function. On the contrary, a membrane that contains the antibiotic valin-gramicidin A retains its original conductivity value even after a temperature decrease below the critical value. Gramicidin A is a peptide with fifteen building units and with a helical structure (see also Figure 88):

$$\begin{array}{l}
\text{HC=O} \\
\quad| \\
\text{L-Val-Gly—L-Ala—D-Leu—L-Ala—D-Val—L-Val—D-Val} \\
\hspace{10.5cm}| \\
\text{NH—L-Trp—D-Leu—L-Trp—D-Leu—L-Trp—D-Leu—L-Trp} \\
\quad| \\
\text{(CH}_2)_2 \\
\quad| \\
\text{OH}
\end{array}$$

(where Val = valine, Gly = glycine, Ala = alanine, Leu = leucine, Trp = tryptophan). The helix has a length such that two molecules of the peptide attached together at their ends just extend beyond the width of the BLM (Figure 87d). Thus, these two helices form a transmembrane channel through which the ions penetrate across the membrane. This channel allows the passage of ions even if the neighbouring lipid molecules are in a crystalline state. The outer surface of the channel must be lipophilic so that it can be placed inside the membrane. Its structure must be sufficiently mobile to enable the shape of the turns of the helix to adjust to the permeating ions. The binding of the ions in the channel must be weaker than for the ionophore because the ion must be released

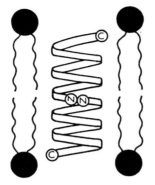

Figure 88. The gramicidin channel consists of two helical molecules that are mutually attached by their ethanol amine endings. (According to V. T. Ivanov)

quickly so that it can pass through the channel. The radius of the turns decides which ions will cross the membrane easily. The selectivity of the gramicidin channel for various monovalent ions obeys the sequence

$$H^+ > NH_4^+ \geqq Cs^+ \geqq Rb^+ > K^+ > Na^+ > Li^+$$

When a small amount of gramicidin A is dissolved in the BLM (the substance is insoluble in water) and the membrane conductivity is measured with a sensitive fast-recording instrument, the time dependence shown in Figure 89 is obtained. The conductivity fluctuates stepwise so that the height of the individual steps is approximately the same. One step corresponds to one channel in the BLM which is open for a definite time and enables the ions to pass across the membrane under the influence of the electric field.

The gramicidin A channel has its origin in a bimolecular reaction of two gramicidin helices and termination of current flow is due to decomposition of the channel. Thus Figure 89 shows the course of incidence of individual

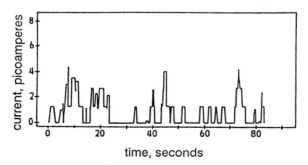

Figure 89. Time dependence of the current flowing across the BLM containing a small amount of gramicidin A. The bathing solutions contain 0.5 M NaCl. Each step corresponds to one opened transmembrane channel, which becomes closed after a definite time period. (According to D. A. Haydon and S. Hladky)

molecular events that are *stochastic* (decided by chance). There are very many natural processes of this kind where similar individual stochastic events are observed (e.g. the formation of a crystal nucleus in electrocrystallization, the formation of adsorption monolayers on electrodes, etc.). The current event alone is, of course, due to the passage of very many ions; in the case of the experiment shown in Figure 89 it is about 10^7 Na^+ per second at 0.1 V imposed on the BLM.

In recent years investigation of gramicidin channels has been in the centre of attention as a model of extremely important channels in biomembranes, particularly of nerve cells (see page 171). These channels were also modelled with the help of helices of synthetic peptides.

The interior of the helix of some channel-forming peptides is partitioned by hydrogen bridges between the CO and NH groups belonging to different amino acid units so that the membrane channel is impermeable. Under the influence of an electric field the dipoles of the carbonyl groups align along the direction of the field and the hydrogen bridges are disrupted, so that the ions can penetrate through the membrane. Therefore the membrane conductivity increases considerably when the electric field strength increases.

Polyene-type antibiotics, like filipin, nystatin or amphotericin B and some other substances, form pores in the membrane. The structure of amphotericin B is as follows:

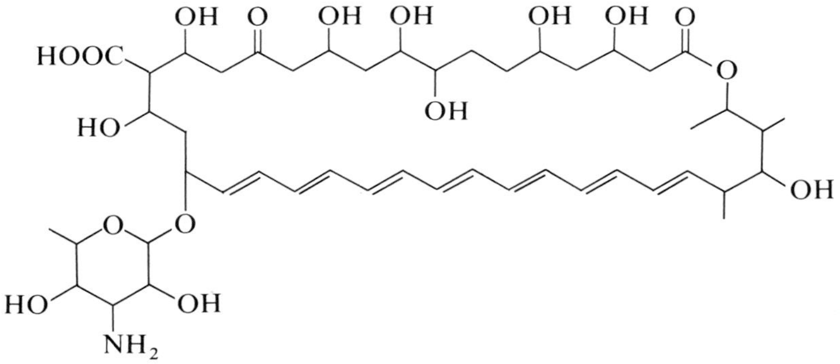

The conductivity of the membrane strongly depends on the concentration of the antibiotic (dependences of $G \sim c^5$ and even $G \sim c^{10}$ have been observed). Thus, the pores in the membrane are formed as a group of several antibiotic molecules. These pores exhibit low selectivity because various cations, water and other non-electrolytes can permeate indiscriminately. However, some selectivity towards anions was observed.

The transport of matter across the membrane has so far depended on only two 'driving forces': the electric field strength and the concentration gradient of the substance transported across the membrane. These two forces can be combined in the gradient of the electrochemical potential of the species concerned (cf. page 69). This is a characteristic property of the *passive transport* which takes place in the direction of the drop (negative gradient) of the electrochemical potential. In biological membranes this kind of transport is far from the only transport phenomenon. The organism has first to form the necessary gradients of electrochemical potential of various substances which can then participate in passive transport or in further chemical processes. To form these gradients energy must be supplied from another source than from the gradient of the electrochemical potential of the transported substance. Thus a new kind of transport, the so-called *active transport* ('uphill transport'), is found in biological systems.

References

1. D. A. Haydon and B. S. Hladky, 'Ion transport across thin lipid membranes: a critical discussion of mechanisms in selected systems', *Q. Rev. Biophys.*, **5**, 187 (1972).
2. A. Pullmann (ed.), *Transport Through Membranes: Carriers, Channels and Pumps*, Kluwer, Dordrecht, 1988.
3. G. Spack (ed.), *Physical Chemistry of Transmembrane Motions*, Elsevier, Amsterdam, 1983.
4. D. W. Urry, *Proc. Nat. Acad. Sci. USA*, **69**, 1610 (1972).

ACTIVE TRANSPORT

The main ionic component in the organism of a frog, sodium ions, are not washed out of its body when the frog stays in water. This phenomenon depends on the active transport of sodium ions in the skin in the direction from the epithelium to the corium, i.e. to the inside of the skin.

In an experiment the skin of the common frog, *Rana temporaria*, representing a membrane is fixed between two compartments of a cell containing identical electrolyte solutions (usually 0.1 M NaCl with additions of KCl, $CaCl_2$ and $NaHCO_3$). In the absence of electric current the electric potential difference between the electrolyte in contact with the outer side of the skin and the other electrolyte is -50 mV. Obviously, an electric potential difference is formed, although the composition of both the solutions is the same. This potential difference can be decreased to zero when an electric current flows in the direction from the outer side to the inner side of the membrane. When the solution on the outer side of the skin is labelled with the radioactive isotope ^{22}Na, it is found that sodium ions are transported from the outer side to the inner side of the skin. Sodium transport occurs in this direction even if the compartment in contact with the inner side of the skin contains a higher sodium ion concentration than the other compartment.

When the temperature of the solution in the cell is increased, the current increases far more than would correspond to mere diffusion or ion migration. When substances inhibiting metabolic processes, e.g. cyanide or ouabain, are added to the cell solution, the current decreases. For example, in the presence of

10^{-4} M ouabain (a digitalis glycoside), whose structure is as follows:

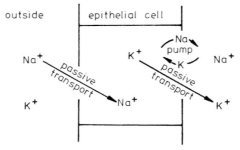

it attains only 5% of the value in the absence of ouabain.

The transfer of Na^+ from the outer solution consists of several steps. The frog skin epithelium is made up of three cell layers. The sides facing the outer solution are permeable to sodium ions (see Figure 90). In the inner side of these cells are regions with built-in molecules of the transport enzyme, *Na, K-ATPase* activated by Mg^{2+}, or *Na^+K^+-pump*. In the membrane of the epithelial cell this protein exposes, towards the cytoplasm, a small cavity in its molecule which can accommodate three sodium ions (Figure 91). At the same time, the protein binds one molecule of adenosine triphosphate (ATP). In the next step the polyphosphate bond between the terminal phosphate unit and the rest of the ATP molecule is broken:

$$ATP + H_3O^+ \overset{ATPase}{\rightleftharpoons} ADP + H_2PO_4^- + H_2O$$

Figure 90. A scheme of sodium transport across frog skin. Larger sized letters indicate higher ion concentration

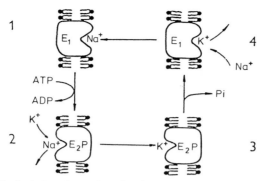

Figure 91. Hypothetical function of Na^+,K^+-ATPase. Position 1: the enzyme binds Na^+. Position 2: after phosphorylation it undergoes a conformational change and transfers Na^+ on the other side of the membrane. Position 3: the enzyme binds K^+. Position 4: after release of the phosphate anion the enzyme acquires the original conformation and transfers K^+ to the other side of the membrane

Here $H_2PO_4^-$ is the phosphate unit which is not set free but is bound to the enzyme molecule (phosphorylation). The energy necessary for this reaction together with the conformation change of the enzyme is supplied by rupture of the energy-rich ('macroergic') polyphosphate bond. As a consequence of this the cavity with bound sodium ions is shifted in the course of a conformation change (i.e. by a change of spatial tertiary structure without changing the sequence of amino acids in the protein) to the other side of the membrane where sodium ions are set free into the intercellular liquid. The vacated cavity now binds two potassium ions. On splitting the phosphate unit from the protein (dephosphorylation) the conformation of the molecule is changed again, in the course of which the cavity with bound potassium ions is displaced to the inner side of the

membrane with subsequent liberation of potassium ions. In this way, sodium ions accumulate in the intercellular liquid and potassium ions within the cell. Ouabain inhibits the dephosphorylation of the enzyme and, therefore, only acts in the case when it is dissolved in the intercellular liquid. Active transport controlled by a chemical reaction rate is called *primary active transport*.

In *secondary active transport* the energy stored in the gradient of electrochemical potential of one substance is used for the transport of another substance against the drop of its electrochemical potential. This case is illustrated by the joint transport of sodium ions (against the gradient of their electrochemical potential, i.e. the transport is passive) and of monosaccharides or amino acids across the epithelial membrane of the small intestine in the direction of the concentration gradient (these are uncharged substances having identical chemical and electrochemical potentials; see page 69).

P. Mitchell classified transport process into three groups. *Uniport* is a type of transport in which only one substance is transported across the membrane. An example is the transport of protons across the purple membrane of the halophilic bacterium, *Halobacterium halobium*, where the energy for transport is provided by light activating the transport protein bacteriorhodopsin. On illumination, its prosthetic group, retinal, is transformed from the all-trans form to the 13-cis form and the protein component is protonated by protons from the cytoplasm. After several milliseconds, retinal is converted back to the trans form and the protons are released into the surrounding medium.

Symport is the interrelated transport of two substances in the same direction, with opposite electrochemical potential gradients—in fact, the identical phenomenon we have already described as secondary active transport.

Antiport is the coupling of two opposite transport processes. A typical example is the transport of Na^+ and K^+, both against their drop in electrochemical potential mediated by Na, K-ATPase and dependent on energy supply from ATP (page 168).

References

1. Page 155, Refs. 1 to 4.
2. A. Kotyk and K. Janáček, *Cell Membrane Transport, Principles and Techniques*, 2nd ed., Plenum Press, New York, 1975.

ELECTROCHEMISTRY OF NERVE EXCITATION

Already at the beginning of the century the fathers of neurophysiology, E. Overton and J. Bernstein, understood the decisive role of sodium and potassium in the activity of the nervous system (later it was shown that, besides these, calcium, acetylcholine and other ions contribute to this activity).

Information transfer in the nerves is based on the transmission of uniform signals, called action potentials, along the nerve fibre. The nerve cell has a body with star-like projections which comprises the cell nucleus and a long tail termed an axon (Figure 92). In some parts the axon is wrapped in a multiple myelin layer so that its membrane only contacts the intercellular liquid in the nodes of

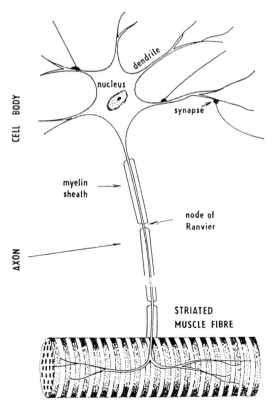

Figure 92. A scheme of a frog motoric nerve cell

Ranvier. The nerve impulses are transferred between nerve fibres in the synapses. An important experimental material is the axons of certain cephalopodes, like squid, or of another sea mollusc *Aplysia*, which are rather thick (with a diameter as large as 1 mm) and are not covered by a myelin layer.

The inside of the axon considerably differs from the intercellular liquid. Thus, for example, the squid axon contains 0.05 M Na^+, 0.4 M K^+, 0.04–0.1 M Cl^-, 0.27 M isethionate and 0.075 M aspartate, while the intercellular liquid comprises 0.46 M Na^+, 0.01 M K^+ and 0.054 M Cl^-.

This difference in concentrations between the outside and the inside of a membrane together with its selective permeability determines the value of the cell resting potential which is characteristic for all living cells. Their inside is charged negatively with respect to the extracellular medium.

When the nerve cell is at rest, a rest membrane potential is established at the axon membrane, $\Delta\varphi_M = \varphi(2) - \varphi(1)$, where $\varphi(2)$ denotes the internal potential of the axon while $\varphi(1)$ is the potential of the intercellular liquid (sometimes $\varphi(2)$ is denoted φ_i and $\varphi(1)$, φ_0). The value of the rest potential is about -70 mV.

A theory of the rest potential of the axon membrane was developed by Goldman, Hodgkin, Huxley and Katz, based on the assumption that the electric field strength in a thin membrane is constant and that the ion transport in the membrane can be described by the Nernst–Planck equation (see page 69). However, this approach does not seem realistic since the ions are transmitted across the membrane through channels that are specific for individual ionic types. In transport through a channel with molecular dimensions it is hardly possible to speak about diffusion because the ions jump across the membrane, which requires overcoming an energy barrier.

Ion jumps will be described by similar equations to the electrode reaction on page 105. Assume that the rest membrane potential is determined solely by transfer of potassium, sodium and chloride ions. For the individual fluxes of the ions it then follows that ($\alpha = \frac{1}{2}$)

$$J_{K^+} = k^0_{K^+}c_{K^+}(1)\exp\left(-F\frac{\Delta\varphi_M}{2RT}\right) - k^0_{K^+}c_{K^+}(2)\exp\left(F\frac{\Delta\varphi_M}{2RT}\right)$$

$$J_{Na^+} = k^0_{Na^+}c_{Na^+}(1)\exp\left(-F\frac{\Delta\varphi_M}{2RT}\right) - k^0_{Na^+}c_{Na^+}(2)\exp\left(F\frac{\Delta\varphi_M}{2RT}\right) \quad (1)$$

$$J_{Cl^-} = k^0_{Cl^-}c_{Cl^-}(1)\exp\left(F\frac{\Delta\varphi_M}{2RT}\right) - k^0_{Cl^-}c_{Cl^-}(2)\exp\left(-F\frac{\Delta\varphi_M}{2RT}\right)$$

Since, at rest, no electric current flows across the membrane, the flux of chloride ions compensates that of sodium and potassium ions:

$$J_{Na^+} + J_{K^+} = J_{Cl^-} \quad (2)$$

Combining equations (1) and (2) yields the expression for the membrane potential:

$$\Delta\varphi_M = \frac{RT}{F}\ln\frac{k^0_{K^+}c_{K^+}(1) + k^0_{Na^+}c_{Na^+}(1) + k^0_{Cl^-}c_{Cl^-}(2)}{k^0_{K^+}c_{K^+}(2) + k^0_{Na^+}c_{Na^+}(2) + k^0_{Cl^-}c_{Cl^-}(1)} \quad (3)$$

Ion transfer is characterized by standard transfer rate constants $k_{K^+}^0$, $k_{Na^+}^0$ and $k_{Cl^-}^0$, which can be identified with the permeabilities for these ions. This simple approach leads to the same result as the treatment of the problem by Hodgkin, Huxley and Katz. Equation (3) satisfactorily explains the value of the rest membrane potential found in experiments, assuming that the permeability of K^+ is larger than that of Na^+ and Cl^-, so that the deviation from the Nernst potential for potassium ions is not very large. On the other hand, the permeabilities for the other ions are not negligible. Consequently, the axon at rest would lose potassium ions and the intracellular sodium concentration would correspondingly grow. This, of course, does not happen because the Na, K-ATPase is again active at the expense of ATP hydrolysis, the enzyme transferring the potassium ions from the intercellular liquid into the axon and the sodium ions in the opposite direction. Since this process is not connected with current flow nor does it influence the membrane potential, it is termed the *electroneutral pump*. In addition, active transport can also occur without 'ion for ion' exchange so that a change in the membrane potential results. This *electrogenic pump* is observed, for example, with muscle fibres when they are kept in a potassium-free, sodium-rich medium for a certain time. They then become loaded with sodium ions through exchange of intercellular potassium for extracellular sodium ions and, after returning to a medium whose composition corresponds to a common intercellular liquid, sodium is extruded from the cells by active transport to a degree such that the membrane potential shifts to more negative values (the cell membrane is *hyperpolarized*). The hyperpolarization is removed by ouabain (cf. page 168).

The basic experiments with electric stimulation of nerve cells were carried out by Cole, Hodgkin and Huxley, who worked with giant squid axons. Their experimental arrangement is shown in Figure 93. The membrane potential is measured by two reference electrodes, e.g. AgCl/Ag, which are attached to the liquids under investigation by micropipettes filled with saline (0.9% NaCl) and are gelified with agar-agar (cf. page 87). One of the micropipettes is immersed in the intercellular liquid close to the axon surface while the other is introduced with the help of a micromanipulator from the side into the axon or impaled into the membrane so that its orifice is placed inside the axon in the intracellular liquid. The auxiliary electrodes are placed either in the positions shown in Figure 93 or one of them is slid into the axon from the side (after cutting off the end of the axon).

As already mentioned, the rest membrane potential is about -70 mV. When the cell is excited with small rectangular current pulses the membrane potential changes, as shown in Figure 94 (the current flowing into the cell is plotted in the diagram). The extent of this change depends on the amount of electricity transferred in the current impulse. With negative current pulses the membrane potential shifts to more negative values—it is hyperpolarized. The current flowing in the opposite direction (a positive current) has a *depolarizing* effect. The potential drops to zero and then increases to positive values. When the pulse exceeds a threshold value a steep potential increase appears. The peak

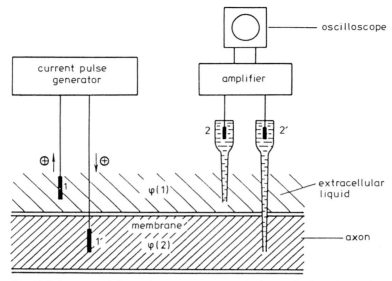

Figure 93. Unit for potential transient measurement during excitation of a squid axon by current pulses from electrodes 1 and 1′; 2 and 2′ are micropipettes. (According to A. L. Hodgkin and A. F. Huxley)

formed is called the *spike* or *action potential*. Its height does not depend on a further increase of the current pulse. Sufficiently large excitation of the membrane results in a large increase in the membrane permeability for sodium ions so that, finally, the membrane potential almost acquires the value of the Nernst potential for sodium ions ($\Delta\varphi_M = +50$ mV). A potential drop to the rest value is accompanied by a temporary influx of sodium ions from the intercellular liquid into the axon.

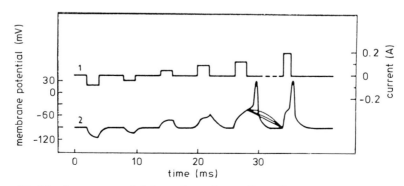

Figure 94. Membrane potential dependence (curve 2) on time course of current pulses (curve 1). The spike sets in at ± 30 s. (According to B. Katz)

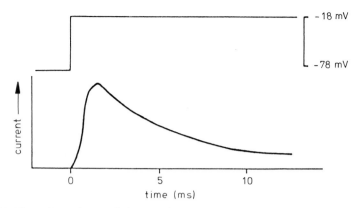

Figure 95. Time dependence of the membrane current. Since the potassium channel is blocked the current corresponds to sodium transport. The upper line represents the time course of the imposed potential difference

By an electrophysiological method of measurement and analysis of current flowing across the membrane at the moment of excitation (the so-called *voltage-clamp* method which is an analogy of the four-electrode potentiostat method; see page 152), Cole, Hodgkin and Huxley succeeded in determining the sequence and magnitude of individual ion currents.* The membrane potential is fixed and the current flowing across the membrane is measured. The resulting current–time dependences have the course shown in Figure 95. Obviously, membrane transport is activated at the start whereas after a definite time it is gradually inhibited. The transport rate typically depends on the composition of both the inner and the outer solution (the cephalopod membrane as much as 1 mm thick can be rinsed with electrolyte solutions without inactivation of its transport mechanisms). The assumption that the current–time dependence is due to current flow through *ion-specific channels* was proved correct by experiments where the axon was affected by various agents. The sodium ion transfer is blocked by tetrodotoxin (page 177) which is a species present in the gonads and the liver of puffer fish, e.g. the fugu of the family Tetraodontidae (for example, *Takufugu rubripes*). This deadly poison (it paralyses all voltage-dependent channels in nerves and muscles and causes death by blocking respiration) is decomposed by heat, so in special restaurants dare-devil Japanese ask for half-fried fugu liver, sometimes with fatal results. The fish does not itself synthesize the poison but feeds on a certain species of alga that contains in its thallus some flagellates which, in their turn, are the actual culprits. During seasons when the

---

* It should be noted that on stimulation of a nerve cell in the *living organism* the electrical parameters of the cell vary while the currents and voltages fluctuate in mutual dependence and in dependence of chemical parameters of the system (channel closing and opening, of effects of stimulators, etc.). In the experiments that are being described now these changes are artificially induced in conditions of *a priori* determined values or time functions of some electrical parameters (current, voltage).

fish lives on other algae, its meat is not poisonous. A similar effect on the sodium channel is exhibited by saxitoxin produced by another flagellate species. The other ion channels have their specific inhibitors; for example the potassium channel is blocked by the tetraethylammonium cation.

A more recent experimental approach to ion membrane channels is the *patch-clamp method*. The orifice of a micropipette filled with an electrolyte solution (with a slightly higher concentration than the bathing solution of the nerve cell) is pressed against the surface of the membrane, causing it to adhere there using slightly reduced pressure. In this way, a patch of the membrane is sucked into the orifice and isolated from the rest of the membrane. In this patch the performance of individual ion channels is recorded as shown in Figure 96. Obviously this course of events is similar to that observed with gramicidin channels (page 164) while the cause of opening and inactivation are different. The potassium channel investigated in Figure 96 by the voltage-clamp method (potentiostatically) is characterized by the probability and length of its opening, both depending on membrane potential.

In contrast with the gramicidin channel, the nerve cell channels are much larger and stable formations. Usually they are glycoproteins, consisting of several subunits. Their hydrophobic region is situated inside the membrane,

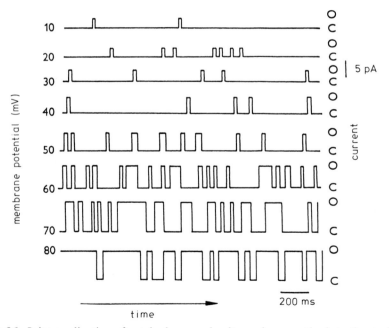

Figure 96. Joint application of patch-clamp and voltage-clamp methods to the study of a single potassium channel present in the membrane of a spinal-cord neuron cultivated in the tissue culture. The values indicated before each curve are potential differences imposed on the membrane. The ion channel is either closed (C) or open (O). (According to B. Hille)

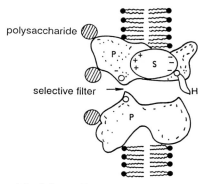

Figure 97. A function model of the sodium channel. P denotes protein, S the potential sensitive sensor and H the gate. The negative sign marks the carboxylate group where the guanidine group of tetrodotoxin can be bound. (According to B. Hille)

while the sugar units stretch out. Ion channels of excitable cells consist of a narrow pore, of a gate that opens and closes the access to the pore and of a sensor that reacts to the stimuli from outside and issues instructions to the gate. The outer stimuli are either a potential change or binding of a specific compound on the sensor. The nerve axon sodium channel was studied in detail (Figure 97). It is a glycoprotein consisting of three subunits, the largest (mol. wt 2.7×10^5) with the pore itself and two smaller ones (mol. wt 3.5×10^4 and 3.3×10^4). The attenuation in the orifice of the pore is a kind of filter (0.5×0.4 nm in size) controlling the entrance of ions with a definite radius. The rate of transport of sodium ions through the channel is considerable: when depolarizing the membrane with $+ 60$ mV a current of approximately 1.5 pA flows through the channel which corresponds to 6×10^6 Na$^+$ ions per second— practically the same value as in the case of the gramicidin A channel. The Na$^+$ channel is only selective but not specific for sodium transport. It shows approximately the same permeability to lithium ions whereas it is roughly ten times lower to potassium. The density of sodium channels strongly varies among different animals, being only 30 μm^{-2} in the case of some marine animals and 330 μm^{-2} in the squid giant axon, reaching 1.2×10^4 μm^{-2} in the mammalian nodes of Ranvier (see Figure 92).

As already mentioned, tetrodotoxin and saxitoxin block the orifice of the sodium channel by binding its guanidine group to the carboxylate group, COO$^-$, present in this orifice:

tetrodotoxin

The activity of the sodium channel is also blocked by other agents. Gate closing is inhibited by toxins isolated from false helleborine (veratridine), from monkshood (aconitine) or from toads (batrachotoxin). The sensor that reacts to membrane potential changes is put out of action by the toxin of sea anemones and some scorpions.

The already mentioned potassium channel (there are many species) is more specific for K^+ than the sodium channel is for Na^+. The potassium channel is almost impermeable to Na^+ while its permeability to ammonium ions is seven times lower and to rubidium is the same as to potassium ions (it should be noted that neither NH_4^+ nor Rb^+ come into contact with nerve fibres in the organism).

In conclusion, it should be stressed that the seemingly smooth curves of the time dependence of potential (Figure 94) or current (Figure 95) consist of innumerable impulses connected with stochastic opening and closing of membrane channels. Formation of the spike (Figure 94) as a result of gradual opening and closing of potassium and sodium channels is schematically shown in Figure 98.

The spike radically changes the electric field in the surroundings of the channel and causes depolarization also in neighbouring channels, thus making possible the transfer of the impulse along the axon. As already pointed out, sodium ions are transferred from the intercellular liquid into the axon during the spike. This gradual formation and disappearance of positive charges corresponds to the flow of positive electric current along the axon. An adequate conductance of thick bare cephalopod axons allows the flow of sufficiently strong currents. In myelinized axons of vertebrates a much larger charge is

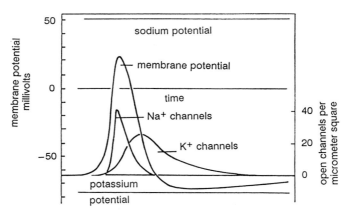

Figure 98. A hypothetic scheme of the time behaviour of the spike linked to the opening and closing of sodium and potassium channels. After longer time intervals a temporary hyperpolarization of the membrane is induced by reversed transport of potassium ions inside the nerve cell. Nernst potential for Na^+ and K^+ are also indicated in the figure. (According to A. L. Hodgkin and A. F. Huxley)

formed (due to the much higher density of sodium channels in the nodes of Ranvier) which moves at a high speed through the much thinner axons than those of cephalopods. The myelin sheath insulates the nerve fibre, impeding in this way the induction of an opposite current in the intercellular liquid which would hinder current flow inside the axon.

References

1. P. F. Baker (ed.), *The Squid Axon*, Current Topics in Membranes and Transport Vol. 22, Academic Press, Orlando, 1984.
2. W. A. Catteral, 'The molecular basis of neuronal excitability', *Science*, **223**, 653 (1984).
3. K. S. Cole, *Membranes, Ions and Impulses*, University of California Press, Berkeley, 1968.
4. R. J. French and R. Horn, 'Sodium channel gating: models, mimics and modifiers', *Ann. Rev. Biophys. Bioengng*, **12**, 319 (1983).
5. B. Hille, *Ionic Channels of Excitable Membranes*, Sinauer, Sunderland, 1984.
6. B. Katz, *Muscle and Synapse*, McGraw-Hill, New York, 1966.
7. T. Narahashi (ed.), *Ion Channels*, Plenum Press, New York.
8. B. Sakmann and E. Neher (eds.), *Single-Channel Recordings*, Plenum Press, New York, 1983.
9. W. D. Stein (ed.), *Ion Channels: Molecular and Physiological Aspects*, Current Topics in Membranes and Transport Vol. 21, Academic Press, Orlando, 1984.
10. C. F. Stevens, 'Biophysical studies in channels', *Science*, **225**, 1346 (1984).

FROM NERVE TO NERVE, FROM NERVE TO MUSCLE

At the end of its passage along the axon the nerve impulse must finally communicate its information to another cell. The connection of the nerve cell with another cell (Figure 92) is called a *synapse* (Greek *synapsis* = unification). A special case of the connection with a muscle cell is termed the *motor endplate* (neuromuscular junction). In those junctions the electric signal is transformed into a chemical signal and, afterwards, back into an electric signal. Figure 99a shows the so-called cholinergic synapse. The end of the axon is terminated by a presynaptic membrane which is separated by a cleft from the postsynaptic membrane. A typical morphological structure in this nerve ending are vesicles (synaptosomes) which contain neurotransmitters (see also page 131). One of the most important neurotransmitters is the acetylcholine cation which is enclosed in the vesicles of the nerve ending in both the synapse and the motor endplate. The presynaptic membrane contains built-in calcium channels (perhaps the most ancient species of a nerve channel). Only a few calcium ions are present inside the cell whereas the intercellular liquid (in the present case, in the synaptic cleft) contains them at an abundant concentration. When the nerve impulse reaches the presynaptic membrane the calcium channels open and calcium ions are bound to the membranes of acetylcholine vesicles. The resulting change of membrane surface charge allows an interaction between them and the presynaptic membrane, resulting in a fusion and in release of acetylcholine into the

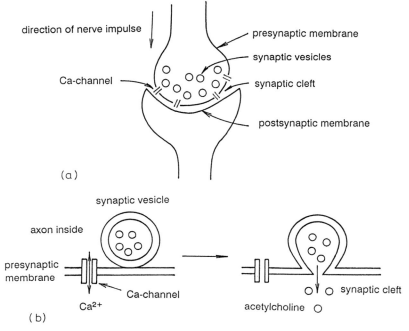

(a)

(b)

Figure 99. The synaptic junction is made possible by neurotransmitters which transform the electric signal to the chemical signal. (a) A scheme of the cholinergic synapse. (b) Calcium ions transported under influence of the spike through the calcium channel induce the acetylcholine vesicles to merge with the synaptic membrane and the pouring out of acetylcholine into the synaptic cleft. (According to L. Stryer)

synaptic cleft (Figure 99b). This phenomenon is called *exocytosis*. Acetylcholine then diffuses to the postsynaptic membrane. These membranes contain the so-called nicotinic acetylcholine receptors, probably the best understood membrane channel. In addition to acetylcholine, it can also be activated by nicotine—hence the term 'nicotinic' (Figure 100). This channel is activated by two acetylcholine molecules. After a short opening period the acetylcholine receptor is deactivated by the enzyme acetylcholinesterase, which breaks down acetylcholine to choline and acetic acid.

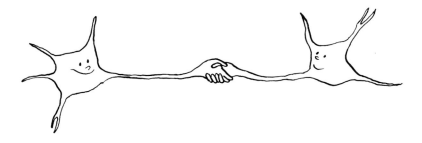

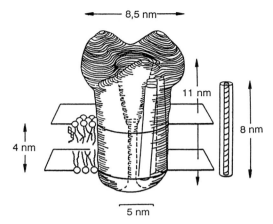

Figure 100. The nicotinic acetylcholine receptor in a membrane. The deciphering of the structure is based on X-ray diffraction and electron microscopy. (According to Kistler and coworkers)

The selectivity of the nicotinic acetylcholine receptor is low as both sodium and potassium ions can permeate with the same ease. It is their transport inside a nerve or muscle cell that is the first stimulus generating an action potential at the postsynaptic membrane or at the motor endplate.

At this stage it is necessary to know some elements of muscle histology. The basic structural unit of a muscle is the muscle cell, the *muscle fibre* (see Figure 101). The muscle fibre is enclosed by the sarcoplasmatic membrane or sarcolemma (Greek *sarx* = flesh, *lemma* = substance). This membrane invaginates into the interior of the fibre by transversal tubules which are filled with the intercellular liquid. The inside of the fibre consists of the actual cytoplasm (sarcoplasm) with inserted mitochondria, sarcoplasmatic reticulum (network) and myofibrils (Greek *myon* = muscle). The myofibrils are the organs of muscle contraction and extension. The sarcoplasmatic reticulum is a membrane system enveloping the myofibrils and leading into the neighbourhood of the tubules through the cisterns.

The genuine excitation of the muscle fibre is contingent on the transfer of sodium and potassium ions across the sarcolemma. The origin of the spike is the same as in the axon, but subsequently the basic role is played by calcium ions. The depolarization of the sarcolemma is accompanied by a decrease in the potential between the transversal tubules and the adjacent parts of the sarcoplasma, resulting in changes in the membrane potential of the sarcoplasmatic reticulum in the neighbourhood. The calcium concentration in the sarcoplasm at rest is lower than 10^{-7} mol dm^{-3} while after excitation it increases to a value of about 10^{-4} mol dm^{-3}. This sudden increase in the calcium concentration activates the myofibrils and elicits their contraction. The myofibrils are composed of parallel thin filaments consisting of the protein actin and of thick filaments built of another protein, myosin. The relative movement of these

182

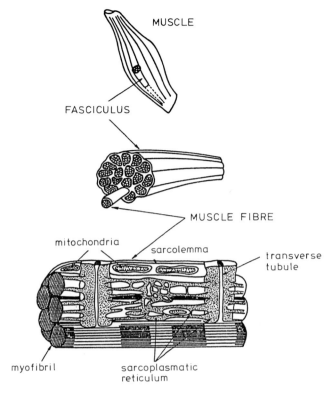

MUSCLE

FASCICULUS

MUSCLE FIBRE

mitochondria sarcolemma transverse tubule

myofibril sarcoplasmatic reticulum

Figure 101. The striated muscle consists of fasciculi that contain individual muscle fibres

filaments, on which muscle contraction is based, depends on the energy supply by hydrolysis of ATP. This movement is hindered by a further protein, troponin, which lies between the thick and thin filaments. The reaction between calcium ions and troponin changes its conformation, resulting in the disappearance of the inhibition and firing of the contraction mechanism. When the contraction stops the sarcoplasmatic calcium is removed into the sarcoplasmatic reticulum by active transport as long as its concentration in the sarcoplasm drops to the original value (about 100 nmol l^{-1}).

References

1. Page 179, Ref. 5.
2. S. Hagiwara and L. Byerly, 'Calcium channel', *Ann. Rev. Neurosci.*, **4**, 69 (1981).
3. S. T. Ohnishi and M. Endo (eds.), *The Mechanism of Gated Calcium Transport Across Biological Membranes*, Academic Press, Orlando, 1982.
4. R. M. Stroud and J. Tiner-Moore, 'Acetylcholine receptor, structure, formation and evolution', *Ann. Rev. Cell Biol.*, **1**, 317 (1985).
5. R. W. Tsien, 'Calcium channels in excitable cell membranes', *Ann. Rev. Physiol.*, **45**, 341 (1983).

MYSTERIOUS ELECTRIC FISH

In this book electric fish have been mentioned in several places (pages 1 and 4). Comparatively strong discharges originate from the electric organ which occupies about four-fifths of the body of this fish. The electric organ consists of several thousand platelet cells (see Figure 102) which contain potassium and sodium ions in similar concentrations as in the nerve cells.

In the course of evolution they were formed from muscle cells. In these cells nicotine acetylcholine receptors are situated in a large amount which makes it possible to isolate several milligrams of the substance from a single fish specimen. This is also the reason why it could be investigated in such detail.

Only the flat upper sides of the platelets are innervated. When the fish is not excited a rest potential builds up around the whole membrane and the electric organ does not function as a voltage source. On excitation a spike sets in on the upper sides of the membranes while the lower folded sides remain at rest. The electric organ then becomes a battery of galvanic cells (with an e.m.f. of about 150 mV each) connected in series. The fish can use the large resulting electric potential difference for stunning or even killing other fish.

Reference

1. R. D. Keynes, 'The generation of electricity by fishes', *Endeavour*, **15**, 215 (1956).

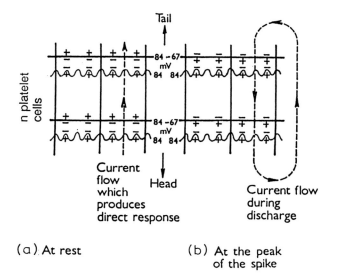

Figure 102. A diagram showing the charging of the platelet cells of the electric eel connected in series. (According to R. A. Keynes)

HOW ENERGY IS SUPPLIED TO AN ORGANISM

As already mentioned on page 132, all the energy necessary for life comes from the Sun, directly or indirectly. The basic energy processes in all living things, included in an interdisciplinary field, bioenergetics, are photosynthesis of carbonaceous fuels, sugars, and conversion of these fuels into more serviceable material, ATP (see Figure 103).

The latter step, cell respiration, is relatively simple and, therefore, will be dealt with first. Here, the Gibbs energy gained by oxidation of carbonaceous substances in cell breathing is accumulated in the reaction of adenosine diphosphate (ADP) with the phosphate ion or, more exactly, the dihydrogen phosphate ion, $H_2PO_4^-$, as the pH of the medium is around 7, to give adenosine triphosphate (ATP; see page 168). This process takes place in membranes of mitochondria, organelles present in cells of the vast majority of multicellular organisms (see Figures 104 and 105).

A number of hypotheses have been put forward concerning the origin of mitochondria. One of them maintains that they originally were symbiotic bacteria somewhat similar to the present genus *Paracoccus* which gradually became adapted to the host eucaryotic cell and lost a part of their genetic outfit which was transferred to the nucleus (the other part remained within the mitochondrion).

The mitochondrion has two membranes. The internal membrane invaginates into the inner space of the organelle (matrix space) through crests (Latin *cristae*). The matrix space contains a considerable amount of various enzymes

185

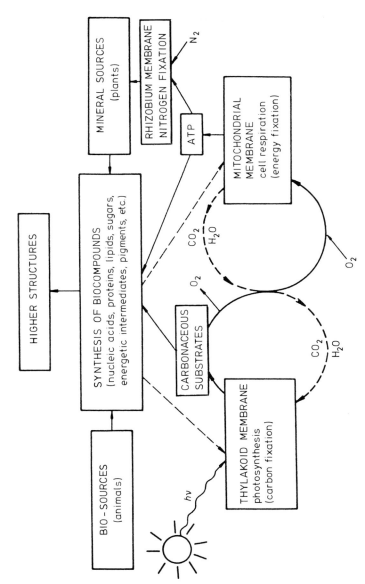

Figure 103. How the energy coming from the Sun is utilized in biological membranes

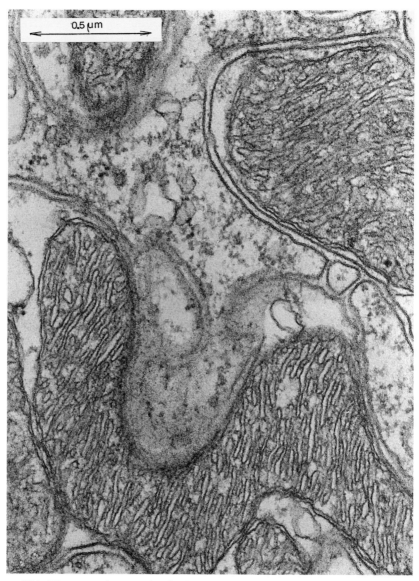

Figure 104. Electron micrograph of mitochondria from rat kidney. On the ultra-thin section the structure of the cytoplasmatic membrane as well as of the mitochondria, including their membranes and the cristae, can be observed. (By courtesy of J. Ludvík)

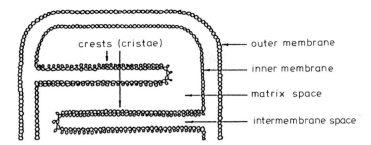

Figure 105. A scheme of the section of a mitochondrion

(Figure 81b) while the space between the internal and the external membrane of the mitochondrion (intermembrane space) contains low-molecular-weight substances including adenine nucleotides (ATP, ADP and AMP, adenosine monophosphate). The oxidation of a carbon substrate by oxygen cannot occur directly because the energy set free would be converted into heat. The process must be carried out in an almost reversible way, with the help of a number of oxidation–reduction systems of low solubility in water which form the *electron transport chain*. The components of the electron transport chain are built into the internal membrane of the mitochondria. They include the nicotinamide adenine dinucleotide (NAD^+/NADH system), several flavoprotein systems, particularly FMN (the flavin mononucleotide), the iron–sulphur proteins (Fe ions are bound in complexes with cysteine of the protein and inorganic sulphide), coenzyme Q (ubiquinone) and a number of cytochromes (for cytochrome c, see page 89 and Figure 46). These redox systems are combined, sometimes with phospholipids and other molecules, into several blocks and ordered across the membrane according to their oxidation–reduction potentials. (The oxidation–reduction potentials of these systems have been measured, if at all, in solution and their actual value in the membrane may differ considerably.) The ordering of redox components enables the oxidation of the carbon substrate to occur 'vectorially', which means that the chemical change is spatially oriented (see Figure 106).

The oxidation of the substrate by oxygen comprises an array of partial oxidation steps, which Mitchell termed loops. Each loop consists of two basic processes, one oriented from the inner membrane surface facing the matrix space to the surface facing the intermembrane space and the other in the opposite direction (Figure 106). In the former process, electrons are transferred together with protons that are extruded into the intermembrane space, while in the latter process only electrons are transported. The first step of the first loop is the reduction of NAD^+ by the carbon substrate SH_2, resulting in a joint transfer of H^+ and electrons. The second step of the last loop is the reduction of oxygen by the cytochrome a_3 system containing copper.

What is the result of this sequence of processes? The electrons are completely utilized for the final reduction of oxygen but the protons accumulate in the

188

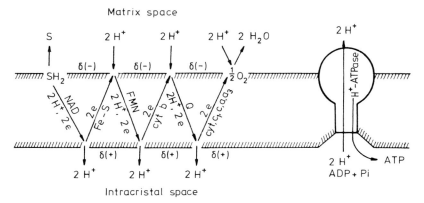

Figure 106. A simplified scheme of electron transfer from the substrate to oxygen via electron-transfer chain enzymes of the inner mitochondrial membrane coupled with hydrogen ion accumulation in the intermembrane space. HS_2 denotes a carbonaceous substrate, for example succinic acid; NAD, nicotinamide adenine dinucleotide; Fe-S, the iron–sulphur protein; FMN, flavin mononucleotide; Q, ubiquinone; cyt, cytochrome; and Pi, the inorganic phosphate ion. Each section represented by two oriented lines, one downwards and the other upwards, forms the Mitchell loop. (According to P. Mitchell)

intermembrane space. Evidently they do not remain there as free ions in solution but they charge the electric double layer at the outer face of the inner mitochondrial membrane, the capacitance of which is large because of extensive folding of its surface. A part of the protons are exchanged for other cations from the intermembrane space which results in a drop of pH in this region. Attempts to separate the increment of proton electrochemical potential, which represents the gained energy, into a chemical component (decrease of pH) and an electric component (increase of membrane potential) seems to be not sufficiently justified although it is found quite frequently in the literature.

How is this accumulated energy utilized? The inner mitochondrial membrane contains a protein, H^+-ATPase, which is able to synthesize ATP using a proton gradient across the membrane. Figure 107 shows a scheme of the structure of this enzyme. A hydrophobic section F_0, which is built into the hydrophobic part of the bilayer, is responsible for the proton transport from the intermembrane space. This part containing several subunits, e.g. OSCP (oligomycin-sensitivity-conferring protein), is linked to the ATP-synthesizing section F_1. The ATP synthesizing function of the enzyme is inhibited by the antibiotic oligomycin. H^+-ATPase acts in a reversible way; i.e. under reversed proton gradient or at low phosphate concentration it converts ATP to ADP and the phosphate ion.

In the presence of various substances the oxidative phosphorylation is uncoupled; i.e. while the carbon substrate is oxidized no ATP is synthesized but it is hydrolysed to ADP and phosphate. The uncouplers of oxidative phosphorylation belong to two groups of ionophores, one transferring protons, the other

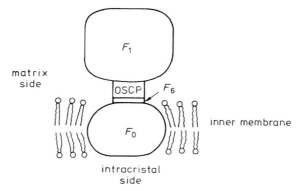

Figure 107. Structure of H^+-ATPase. (According to A. E. Senior)

alkali metal ions. The first group includes some acid anions, such as 2,4-dinitrophenolate:

Because of spreading of the negative charge around the whole structure, the anion can penetrate the lipidic part of the mitochondrial membrane in the direction of increasing electric potential. When it reaches the intermembrane space with a high hydrogen ion concentration, it is transformed to a non-dissociated form of the acid which is then transported back to the matrix space. Here it splits off the proton which has been acquired in the intermembrane space.

Both the transport of the anion into the intermembrane space and of the acid are passive and therefore decrease both the membrane potential and the hydrogen activity in the intermembrane space. In this way the energy gained in the electron-transfer chain is dissipated as heat.

The second group of substances are the macrocyclic complex-formers, such as valinomycin. With this species the most pronounced uncoupling occurs in the presence of potassium ions (cf. its effect on the potential of the ion-selective electrode on page 144). As a result of its lipophilicity the electrically uncharged ionophore freely diffuses in the inner membrane of the mitochondrion. On the intermembrane side of the membrane it combines with the potassium ion. The complex is then transported passively to the matrix space. This process, which in fact simulates the proton transport, decreases the membrane potential. The elucidation of the uncoupling of oxidative phosphorylation represents a notable support for the Mitchell theory.

Cell respiration is a basic step through which energy is supplied to the organism, but an absolutely necessary source for this oxidation process is the 'fuel', i.e. the carbon compounds, primarily sugars, formed in photosynthesis, the energy of sunlight being the primary energy source for all living organisms. In green plants the light energy is supplied to an oxidation–reduction reaction between sufficiently abundant components, water and carbon dioxide. This reaction occurs according to the equation

$$6\,CO_2 + 6\,H_2O \xrightarrow[\text{chlorophyll}]{hv} C_6H_{12}O_6 + 6\,O_2$$

while in phototrophic bacteria the reaction

$$6\,CO_2 + 12\,H_2S \xrightarrow[\text{bacteriochlorophyll}]{hv} C_6H_{12}O_6 + 12\,S + 6\,H_2O$$

takes place. The latter reaction cannot, of course, become a universal photosynthetic process because hydrogen sulphide is unstable, is not present in sufficient amounts and is produced mainly by biological processes. The only substance that can be exploited for carbon dioxide reduction and is present in the biosphere independently of biological processes is water. The most important catalysts of photosynthesis are various kinds of porphyrin complexes of magnesium, the chlorophylls. The halophilic (Greek *hals* = salt) archaebacterium *Halobacterium halobium* contains the protein bacteriorhodopsin (Greek *rhodos* = purple, *opsis* = sight), which is structurally related to the visual pigment, rhodopsin, as a photosynthetic substance (cf. page 170).

Photosynthesis in green plants proceeds through two basic processes. The dark process (the Calvin cycle) is the reduction of carbon dioxide by a strong reductant, nicotinamide adenine dinucleotidephosphate, with a supply of energy gained by the hydrolysis of ATP to ADP:

$$6\,CO_2 + 12\,NADPH_2 + 6\,H_2O \longrightarrow C_6H_{12}O_6 + 12\,NADP$$

$$ATP \longrightarrow ADP + P_{inorg}$$

The light process is a typical membrane process. The membrane of thylakoids, which are structures typical of the chloroplast (see Figure 108), contains chlorophyll (or, more exactly, an integrated system of several chlorophyll-type pigments and carotenes) and an enzyme system that makes possible the accumulation of the energy of sunlight in the form of nicotinamide adenine dinucleotidephosphate and ATP (this kind of formation of ATP from ATP and a phosphate ion is called photophosphorylation and is the same as described with mitochondrial formation of ATP).

As shown in Figure 109, the light process occurs in two photosystems. On absorption of a light quantum by the chlorophyll of photosystem II, charge

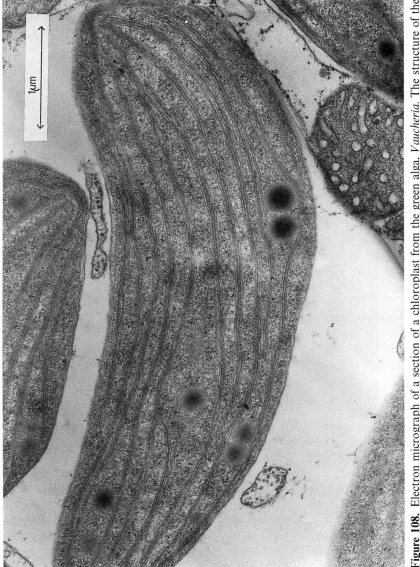

Figure 108. Electron micrograph of a section of a chloroplast from the green alga, *Vaucheria*. The structure of the basic units of the chloroplast, thylakoids, is easily perceptible. (By courtesy of J. Ludvik)

192

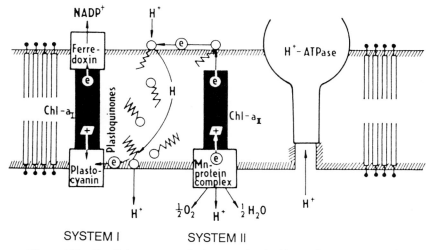

Figure 109. The light process of photosynthesis. (According to H. T. Witt)

separation sets in, i.e. an electron–hole pair is formed. The hole is annihilated by an electron from the enzyme containing manganese in its prosthetic group. The extraction of a total of four electrons (connected with absorption of four light quanta by photosystem II) is necessary for the evolution of an oxygen molecule. This process requires the concerted action of four manganese-containing sites in the enzyme (each containing a pair of manganese atoms) where manganese is assumed to occur as Mn^{2+}, Mn^{3+} and Mn^{4+} and where hydroxide ions bound in complexes with manganese are transformed to bound superoxide or hydroperoxide groups from which molecular oxygen is set free. At the same time, protons cross the inner side of the thylakoid membrane and accumulate inside the thylakoid.

The energy of the electron formed in photosystem II is not high enough to reduce NADP so another light quantum is required. To accomplish this the electron reacts first with hydrophobic quinone, the plastoquinone, on the outer surface of the thylakoid membrane. Its reduced form accepts two protons from the outer solution. Both the electrons and protons are transferred to further plastoquinone molecules and the electrons finally arrive at the enzyme, plastocyanin, from where they annihilate the holes formed in photosystem I after absorption of light. At the same time, the protons originating in the outer solution are transferred into the thylakoid. The high-energy electron set free in photosystem I reduces the Fe–S protein, ferredoxin, and through a chain of several other enzymes attacks the oxidized nicotinamide, adenine dinucleotide-phosphate. A part of the energy gained is stored in the form of increased electrochemical potential of the hydrogen ions in the interior of the thylakoid. This energy is then supplied to the H^+-ATPase. In contrast to cell respiration, where the energy from the oxidation of a carbon substrate is accumulated in a

single 'macroergic' compound, ATP, in photosynthesis two energy-rich compounds are formed, $NADPH_2$ and ATP, whose energy is then consumed for the final sugar synthesis. The efficiency of the photosynthetic process alone, without losses due to imperfect light absorption by the green plant, is about 34%, which is fairly good compared with solid-state or electrochemical photovoltaic devices. On the other hand, this is far from the perfect energy transformation, a physicochemical property that laymen ascribe to biological systems. Nature does not need total accumulation of solar energy; the environment of the plants must also be heated by sunlight.

References

1. W. A. Cramer and D. B. Knaff, *Energy Transduction in Biological Membranes, A Textbook of Bioenergetics*, Springer-Verlag, Berlin, 1989.
2. P. Mitchell, 'Davy's electrochemistry—Nature's protochemistry', *Chemistry in Britain*, **17**, 14 (1981).
3. D. G. Nicholls, *Bioenergetics, An Introduction to the Chemismotic Theory*, Academic Press, Orlando, 1982.
4. A. Tzagoloff, *Mitochondria*, Plenum Press, New York, 1982.
5. W. R. Briggs (ed.), *Photosynthesis*, A. R. Liss, New York, 1990.
6. Govindjee (ed.), *Photosynthesis*, Vols. 1 and 2, Academic Press, New York, 1982.
7. C. H. Foyer, *Photosynthesis*, John Wiley & Sons, Chichester, 1984.

Index